Cram101 Textbook Outlines to accompany:

Physical Chemistry: For the Biosciences

Raymond Chang, 1st Edition

A Cram101 Inc. publication (c) 2011.

PRACTICE EXAMS.

Get all of the self-teaching practice exams for each chapter of this textbook at
www.Cram101.com and ace the tests. Here is an example:

Chapter 1

Physical Chemistry: For the Biosciences
Raymond Chang, 1st Edition,
All Material Written and Prepared by Cram101

I WANT A BETTER GRADE.   Items 1 - 50 of 100. ▶

1 _____ is the iron-containing oxygen-transport metalloprotein in the red blood cells of vertebrates, and the tissues
of some invertebrates.

In mammals, the protein makes up about 97% of the red blood cell"s dry content, and around 35% of the total content .
_____ transports oxygen from the lungs or gills to the rest of the body where it releases the oxygen for cell use.

- ◯ Hemoglobin
- ◯ Half-life
- ◯ Haber process
- ◯ Halochromic material

2 _____ is the force per unit area applied in a direction perpendicular to the surface of an object. Gauge _____
is the _____ relative to the local atmospheric or ambient _____.

_____ is an effect which occurs when a force is applied on a surface.

- ◯ Pressure
- ◯ Packaged terminal air conditioner
- ◯ P. R. Mallory and Co Inc
- ◯ Paint

3 The standard _____ is an international reference pressure defined as 101,325 Pa and formerly used as unit of
pressure (symbol: atm.) For practical purposes it has been replaced by the bar which is 100,000 Pa. The difference of
about 1% is not significant for many applications, and is within the error range of common pressure gauges.

- ◯ Atmosphere
- ◯ ABCN
- ◯ Ab initio multiple spawning
- ◯ ABINIT

You get a 50% discount for the online exams. Go to **Cram101.com**, click Sign Up at
the top of the screen, and enter DK73DW6079 in the promo code box on the
registration screen. Access to Cram101.com is $4.95 per month, cancel at any time.

With Cram101.com online, you also have access to extensive reference material.

You will nail those essays and papers. Here is an example from a Cram101 Biology text:

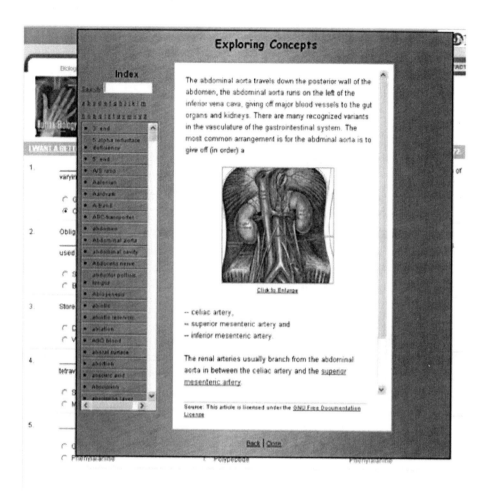

Learning System

Cram101 Textbook Outlines is a learning system. The notes in this book are the highlights of your textbook, you will never have to highlight a book again.

How to use this book. Take this book to class, it is your notebook for the lecture. The notes and highlights on the left hand side of the pages follow the outline and order of the textbook. All you have to do is follow along while your instructor presents the lecture. Circle the items emphasized in class and add other important information on the right side. With Cram101 Textbook Outlines you'll spend less time writing and more time listening. Learning becomes more efficient.

Cram101.com Online

Increase your studying efficiency by using Cram101.com's practice tests and online reference material. It is the perfect complement to Cram101 Textbook Outlines. Use self-teaching matching tests or simulate in-class testing with comprehensive multiple choice tests, or simply use Cram's true and false tests for quick review. Cram101.com even allows you to enter your in-class notes for an integrated studying format combining the textbook notes with your class notes.

Visit **www.Cram101.com**, click Sign Up at the top of the screen, and enter **DK73DW6079** in the promo code box on the registration screen. Access to www.Cram101.com is normally $9.95 per month, but because you have purchased this book, your access fee is only $4.95 per month. Sign up and stop highlighting textbooks forever.

Physical Chemistry: For the Biosciences
Raymond Chang, 1st

CONTENTS

Hemoglobin	Hemoglobin is the iron-containing oxygen-transport metalloprotein in the red blood cells of vertebrates, and the tissues of some invertebrates.
	In mammals, the protein makes up about 97% of the red blood cell"s dry content, and around 35% of the total content . Hemoglobin transports oxygen from the lungs or gills to the rest of the body where it releases the oxygen for cell use.
Pressure	Pressure is the force per unit area applied in a direction perpendicular to the surface of an object. Gauge Pressure is the Pressure relative to the local atmospheric or ambient Pressure.
	Pressure is an effect which occurs when a force is applied on a surface.
Atmosphere	The standard Atmosphere is an international reference pressure defined as 101,325 Pa and formerly used as unit of pressure (symbol: atm.) For practical purposes it has been replaced by the bar which is 100,000 Pa. The difference of about 1% is not significant for many applications, and is within the error range of common pressure gauges.

Closed system	A Closed system is a system in the state of being isolated from its surrounding environment. The term often refers to an idealized system in which closure is perfect. In reality no system can be completely closed; there are only varying degrees of closure.
Isolated system	In the natural sciences an Isolated system, as contrasted with a open system, is a physical system that does not interact with its surroundings. It obeys a number of conservation laws: its total energy and mass stay constant. They cannot enter or exit, but can only move around inside.
Gas	In physics, a Gas is a state of matter, consisting of a collection of particles (molecules, atoms, ions, electrons, etc.) without a definite shape or volume that are in more or less random motion. Due to the electronic nature of the aforementioned particles, a "force field" is present throughout the space around them.
Ideal gas	An Ideal gas is a theoretical gas composed of a set of randomly-moving point particles that interact only through elastic collisions. The Ideal gas concept is useful because it obeys the Ideal gas law, a simplified equation of state, and is amenable to analysis under statistical mechanics. At normal ambient conditions such as standard temperature and pressure, most real gases behave qualitatively like an Ideal gas.
Thermodynamic	In physics, Thermodynamic s ">power") is the study of the conversion of energy into work and heat and its relation to macroscopic variables such as temperature,volume and pressure. Its underpinnings, based upon statistical predictions of the collective motion of particles from their microscopic behavior, is the field of statistical Thermodynamic s, a branch of statistical mechanics. Historically, Thermodynamic s developed out of need to increase the efficiency of early steam engines.
Zeroth law of thermodynamics	The Zeroth law of thermodynamics is a generalization about the thermal equilibrium among bodies in contact. It results from the definition and properties of temperature. It can be stated mathematically as: $$A \sim B \wedge B \sim C \Rightarrow A \sim C$$ A system is said to be in thermal equilibrium when its temperature does not change over time.
Equation of state	In physics and thermodynamics, an Equation of state is a relation between state variables. More specifically, an Equation of state is a thermodynamic equation describing the state of matter under a given set of physical conditions. It is a constitutive equation which provides a mathematical relationship between two or more state functions associated with the matter, such as its temperature, pressure, volume, or internal energy.

Gas constant

The Gas constant is a physical constant which is featured in a large number of fundamental equations in the physical sciences, such as the ideal gas law and the Nernst equation. It is equivalent to the Boltzmann constant, but expressed in units of energy (i.e. the pressure-volume product) per kelvin per mole (rather than energy per kelvin per particle.)

Its value is

$$R = 8.314\,472(15)\,\frac{\text{J}}{\text{K\,mol}}.$$

The two digits in parentheses are the uncertainty (standard deviation) in the last two digits of the value.

Pressure

Pressure is the force per unit area applied in a direction perpendicular to the surface of an object. Gauge Pressure is the Pressure relative to the local atmospheric or ambient Pressure.

Pressure is an effect which occurs when a force is applied on a surface.

Activation

Activation in (bio-)chemical sciences generally refers to the process whereby something is prepared or excited for a subsequent reaction.

In chemistry, Activation of molecules is required for a chemical reaction to occur. The phrase energy of Activation refers to the energy the reactants must acquire before they can successfully react with each other to produce the products, that is, to reach the transition state.

Phase

In the physical sciences, a Phase is a region of space (a thermodynamic system), throughout which all physical properties of a material are essentially uniform. Examples of physical properties include density, index of refraction, and chemical composition. A simple description is that a Phase is a region of material that is chemically uniform, physically distinct, and (often) mechanically separable.

Partial pressure

In a mixture of ideal gases, each gas has a Partial pressure which is the pressure which the gas would have if it alone occupied the volume. The total pressure of a gas mixture is the sum of the Partial pressure s of each individual gas in the mixture.

In chemistry, the Partial pressure of a gas in a mixture of gases is defined as above.

Mole fraction

In chemistry, Mole fraction x (also, and more correctly, known as the amount fraction) is a way of expressing the composition of a mixture. The Mole fraction of each component i is defined as its amount of substance n_i divided by the total amount of substance in the system, n

$$x_i \stackrel{\text{def}}{=} \frac{n_i}{n}$$

where

$$n = \sum_i n_i$$

The sum is over all components, including the solvent in the case of a chemical solution. As an example, if a mixture is obtained by dissolving 10 moles of sucrose in 90 moles of water, the Mole fraction of sucrose in that mixture is 0.1.

Compressibility

In thermodynamics and fluid mechanics, Compressibility is a measure of the relative volume change of a fluid or solid as a response to a pressure (or mean stress) change.

$$\beta = -\frac{1}{V}\frac{\partial V}{\partial p}$$

where V is volume and p is pressure. The above statement is incomplete, because for any object or system the magnitude of the Compressibility depends strongly on whether the process is adiabatic or isothermal.

Compressibility factor

The Compressibility factor is a useful thermodynamic property for modifying the ideal gas law to account for the real gas behaviour. In general, deviations from ideal behavior become more significant the closer a gas is to a phase change, the lower the temperature or the larger the pressure. Compressibility factor values are usually obtained by calculation from equations of state (EOS), such as the virial equation which take compound specific empirical constants as input.

Real gas

Real gas as opposed to a Perfect or Ideal Gas, effects refers to an assumption base where the following are taken into account:

- Compressibility effects
- Variable heat capacity
- Van der Waals forces
- Non-equilibrium thermodynamic effects
- Issues with molecular dissociation and elementary reactions with variable composition.

For most applications, such a detailed analysis is "over-kill" and the ideal gas approximation is used. Real-gas models have to be used near condensation point of gases, near critical point, at very high pressures, and in several other less usual cases.

Real gas es are often modeled by taking into account their molar weight and molar volume

$$RT = (P + \frac{a}{V_m^2})(V_m - b)$$

Where P is the pressure, T is the temperature, R the ideal gas constant, and V_m the molar volume. a and b are parameters that are determined empirically for each gas, but are sometimes estimated from their critical temperature (T_c) and critical pressure (P_c) using these relations:

$$a = \frac{27 R^2 T_c^2}{64 P_c}$$

$$b = \frac{RT_c}{8 P_c}$$

The Redlich-Kwong equation is another two-parameters equation that is used to modelize Real gas es.

Van der Waals Equation	The Van der Waals equation is an equation of state for a fluid composed of particles that have a non-zero size and a pairwise attractive inter-particle force (such as the van der Waals force.) It was derived by Johannes Diderik van der Waals in 1873, based on a modification of the ideal gas law, who received the Nobel prize in 1910 for "his work on the equation of state for gases and liquids". The equation approximates the behavior of real fluids, taking into account the nonzero size of molecules and the attraction between them.
Hemoglobin	Hemoglobin is the iron-containing oxygen-transport metalloprotein in the red blood cells of vertebrates, and the tissues of some invertebrates.
	In mammals, the protein makes up about 97% of the red blood cell"s dry content, and around 35% of the total content . Hemoglobin transports oxygen from the lungs or gills to the rest of the body where it releases the oxygen for cell use.
Critical point	In physical chemistry, thermodynamics, chemistry and condensed matter physics, a Critical point specifies the conditions at which a phase boundary ceases to exist. There are multiple types of critical points such as vapor-liquid critical points and liquid-liquid critical points. A plot of typical polymer solution phase behavior including two critical points: an LCST and a UCST.
	The liquid-liquid Critical point of a solution denotes the limit of the two-phase region of the phase diagram.

Vapor pressure	Vapor pressure, is the pressure of a vapor in equilibrium with its non-vapor phases. All liquids and solids have a tendency to evaporate to a gaseous form, and all gases have a tendency to condense back into their original form At any given temperature, for a particular substance, there is a pressure at which the gas of that substance is in dynamic equilibrium with its liquid or solid forms.
Vapor	A Vapor or vapour is a substance in the gas phase at a temperature lower than its critical temperature. This means that the Vapor can be condensed to a liquid or to a solid by increasing its pressure, without reducing the temperature. For example, water has a critical temperature of 374 °C (or 647 K) which is the highest temperature at which liquid water can exist.
Kinetic theory	Kinetic theory attempts to explain macroscopic properties of gases, such as pressure, temperature by considering their molecular composition and motion. Essentially, the theory posits that pressure is due not to static repulsion between molecules, as was Isaac Newton"s conjecture, but due to collisions between molecules moving at different velocities. Kinetic theory is also known as the Kinetic-Molecular Theory or the Collision Theory or the Kinetic-Molecular Theory of Gases.
Supercritical fluid	A Supercritical fluid is any substance at a temperature and pressure above its critical point. It can diffuse through solids like a gas, and dissolve materials like a liquid. Additionally, close to the critical point, small changes in pressure or temperature result in large changes in density, allowing many properties to be "tuned".
Boltzmann constant	The Boltzmann constant is the physical constant relating energy at the particle level with temperature observed at the bulk level. It is the gas constant R divided by the Avogadro constant N_A: $$k = \frac{R}{N_A}.$$ It has the same units as entropy. It is named after the Austrian physicist Ludwig Boltzmann.
Activation energy	In chemistry, Activation energy is a term introduced in 1889 by the Swedish scientist Svante Arrhenius, that is defined as the energy that must be overcome in order for a chemical reaction to occur. Arrhenius" research was a follow up of the theories of reaction rate by Serbian physicist Nebojsa Lekovic. Activation energy may also be defined as the minimum energy required to start a chemical reaction.

Effusion

In chemistry, Effusion is the process in which individual molecules flow through a hole without collisions between molecules. This occurs if the diameter of the hole is considerably smaller than the mean free path of the molecules. According to Graham"s law, the rate at which gases effuse (i.e., how many molecules pass through the hole per second) is dependent on their molecular weight; gases with a lower molecular weight effuse more quickly than gases with a higher molecular weight.

Gas	In physics, a Gas is a state of matter, consisting of a collection of particles (molecules, atoms, ions, electrons, etc.) without a definite shape or volume that are in more or less random motion.
	Due to the electronic nature of the aforementioned particles, a "force field" is present throughout the space around them.
First law of thermodynamics	The First law of thermodynamics, an expression of the principle of conservation of energy, states that energy can be transformed (changed from one form to another), but it can neither be created nor destroyed. Alternatively:
	The First law of thermodynamics basically states that a thermodynamic system can store or hold energy and this energy, called internal energy, is conserved. Heat is a process by which energy is either added to a system from a high-temperature source or lost from a system to a low-temperature sink.
Thermodynamic	In physics, Thermodynamic s ">power") is the study of the conversion of energy into work and heat and its relation to macroscopic variables such as temperature,volume and pressure. Its underpinnings, based upon statistical predictions of the collective motion of particles from their microscopic behavior, is the field of statistical Thermodynamic s, a branch of statistical mechanics. Historically, Thermodynamic s developed out of need to increase the efficiency of early steam engines.
Internal energy	In thermodynamics, the Internal energy of a thermodynamic system denoted by U is the total of the kinetic energy due to the motion of molecules (translational, rotational, vibrational) and the potential energy associated with the vibrational and electric energy of atoms within molecules or crystals. It includes the energy in all of the chemical bonds, and the energy of the free, conduction electrons in metals.
	One can also calculate the Internal energy of electromagnetic or blackbody radiation.
Enthalpy	In thermodynamics and molecular chemistry, the Enthalpy is a thermodynamic property of a fluid. It can be used to calculate the heat transfer during a quasistatic process taking place in a closed thermodynamic system under constant pressure. Enthalpy H is an arbitrary concept but the Enthalpy change ΔH is more useful because it is equal to the change in the internal energy of the system, plus the work that the system has done on its surroundings.
Heat	In physics and thermodynamics, Heat is the process of energy transfer from one body or system to another due to a difference in temperature. In thermodynamics, the quantity TdS is used as a representative measure of the (inexact) Heat differential δQ, which is the absolute temperature of an object multiplied by the differential quantity of a system"s entropy measured at the boundary of the object.
	A related term is thermal energy, loosely defined as the energy of a body that increases with its temperature.

Electron configuration	In atomic physics and quantum chemistry, Electron configuration is the arrangement of electrons of an atom, a molecule, or other physical structure. It concerns the way electrons can be distributed in the orbitals of the given system (atomic or molecular for instance.) Like other elementary particles, the electron is subject to the laws of quantum mechanics, and exhibits both particle-like and wave-like nature.
Isothermal process	An Isothermal process is a change in which the temperature of the system stays constant: $\Delta T = 0$. This typically occurs when a system is in contact with an outside thermal reservoir (heat bath), and the change occurs slowly enough to allow the system to continually adjust to the temperature of the reservoir through heat exchange. An alternative special case in which a system exchanges no heat with its surroundings ($Q = 0$) is called an adiabatic process.
Adiabatic process	In thermodynamics, an Adiabatic process or an isocaloric process is a thermodynamic process in which no heat is transferred to or from the working fluid. The term "adiabatic" literally means impassable, coming from the Greek roots á¼€- , διá½°- ("through"), and βαá¿–νειν ("to pass"); this etymology corresponds here to an absence of heat transfer. Conversely, a process that involves heat transfer (addition or loss of heat to the surroundings) is generally called diabatic.
Activation	Activation in (bio-)chemical sciences generally refers to the process whereby something is prepared or excited for a subsequent reaction. In chemistry, Activation of molecules is required for a chemical reaction to occur. The phrase energy of Activation refers to the energy the reactants must acquire before they can successfully react with each other to produce the products, that is, to reach the transition state.
Phase	In the physical sciences, a Phase is a region of space (a thermodynamic system), throughout which all physical properties of a material are essentially uniform. Examples of physical properties include density, index of refraction, and chemical composition. A simple description is that a Phase is a region of material that is chemically uniform, physically distinct, and (often) mechanically separable.
Endothermic	In thermodynamics, the word Endothermic "within-heating" describes a process or reaction that absorbs energy in the form of heat. Its etymology stems from the Greek prefix endo-, meaning "inside" and the Greek suffix -thermic, meaning "to heat". The opposite of an Endothermic process is an exothermic process, one that releases energy in the form of heat.
Exothermic	In thermodynamics, the term Exothermic describes a process or reaction that releases energy usually in the form of heat, but also in form of light (e.g. a spark, flame, or explosion), electricity (e.g. a battery), or sound. Its etymology stems from the Greek prefix ex- and the Greek word thermein (meaning "to heat".) The term Exothermic was first coined by Marcellin Berthelot.
Exothermic reaction	An Exothermic reaction is a chemical reaction that releases energy in the form of heat. It is the opposite of an endothermic reaction. Expressed in a chemical equation: reactants → products + energy

An Exothermic reaction is a chemical reaction that is accompanied by the release of heat.

Thermochemistry

In thermodynamics and physical chemistry, Thermochemistry is the study of the energy evolved or absorbed in chemical reactions and any physical transformations, such as melting and boiling. Thermochemistry, generally, is concerned with the energy exchange accompanying transformations, such as mixing, phase transitions, chemical reactions, and including calculations of such quantities as the heat capacity, heat of combustion, heat of formation, enthalpy, and free energy.

Thermochemistry rests on two generalizations:

1. Lavoisier and Laplace"s law (1782): the heat exchange accompanying a transformation is equal and opposite to the heat exchange accompanying the reverse transformation.
2. Hess"s law (1840): the heat exchange accompanying a transformation is the same whether the process occurs in one or both steps

Both of these statements preceded the first law of thermodynamics (1850) and helped in its formulation.

Lavoisier, Laplace and Hess also investigated specific heat and latent heat, although it was Joseph Black who made the most important contributions to the development of latent energy changes.

Dissociation

Dissociation in chemistry and biochemistry is a general process in which ionic compounds (complexes, molecules ions usually in a reversible manner. When a Bronsted-Lowry acid is put in water, a covalent bond between an electronegative atom and a hydrogen atom is broken by heterolytic fission, which gives a proton and a negative ion. Dissociation is the opposite of association and recombination.

Entropy	Entropy is a concept applied across physics, information theory, mathematics and other branches of science and engineering. The following definition is shared across all these fields: $$S = -k \sum_i P_i \ln P_i$$ where S is the conventional symbol for Entropy. The sum runs over all microstates consistent with the given macrostate and P_i is the probability of the ith microstate.
Gas	In physics, a Gas is a state of matter, consisting of a collection of particles (molecules, atoms, ions, electrons, etc.) without a definite shape or volume that are in more or less random motion. Due to the electronic nature of the aforementioned particles, a "force field" is present throughout the space around them.
Thermodynamic	In physics, Thermodynamic s ">power") is the study of the conversion of energy into work and heat and its relation to macroscopic variables such as temperature,volume and pressure. Its underpinnings, based upon statistical predictions of the collective motion of particles from their microscopic behavior, is the field of statistical Thermodynamic s, a branch of statistical mechanics. Historically, Thermodynamic s developed out of need to increase the efficiency of early steam engines.
Heat	In physics and thermodynamics, Heat is the process of energy transfer from one body or system to another due to a difference in temperature. In thermodynamics, the quantity TdS is used as a representative measure of the (inexact) Heat differential δQ, which is the absolute temperature of an object multiplied by the differential quantity of a system"s entropy measured at the boundary of the object. A related term is thermal energy, loosely defined as the energy of a body that increases with its temperature.
Heat engine	In engineering and thermodynamics, a Heat engine performs the conversion of heat energy to mechanical work by exploiting the temperature gradient between a hot "source" and a cold "sink". Heat is transferred from the source, through the "working body" of the engine, to the sink, and in this process some of the heat is converted into work by exploiting the properties of a working substance (usually a gas or liquid.) Figure 1: Heat engine diagram Heat engine s are often confused with the cycles they attempt to mimic.
Second law of thermodynamics	The Second law of thermodynamics is an expression of the universal principle of increasing entropy, stating that the entropy of an isolated system which is not in equilibrium will tend to increase over time, approaching a maximum value at equilibrium.

The origin of the second law can be traced to French physicist Sadi Carnot"s 1824 paper Reflections on the Motive Power of Fire, which presented the view that motive power is due to the flow of caloric (heat) from a hot to cold body (working substance.) In simple terms, the second law is an expression of the fact that over time, ignoring the effects of self-gravity, differences in temperature, pressure, and density tend to even out in a physical system that is isolated from the outside world.

Phase	In the physical sciences, a Phase is a region of space (a thermodynamic system), throughout which all physical properties of a material are essentially uniform. Examples of physical properties include density, index of refraction, and chemical composition. A simple description is that a Phase is a region of material that is chemically uniform, physically distinct, and (often) mechanically separable.
Vaporization	Vaporization of an element or compound is a phase transition from the liquid phase to gas phase. There are two types of Vaporization: evaporation and boiling. This diagram shows the nomenclature for the different phase transitions. Evaporation is a phase transition from the liquid phase to gas phase that occurs at temperatures below the boiling temperature at a given pressure.
Third law of thermodynamics	The Third law of thermodynamics is a statistical law of nature regarding entropy and the impossibility of reaching absolute zero of temperature The essence of the postulate is that the entropy of the given system near absolute zero depends only on the temperature (i.e. tends to a constant independently of the other parameters.)
Enthalpy	In thermodynamics and molecular chemistry, the Enthalpy is a thermodynamic property of a fluid. It can be used to calculate the heat transfer during a quasistatic process taking place in a closed thermodynamic system under constant pressure. Enthalpy H is an arbitrary concept but the Enthalpy change ΔH is more useful because it is equal to the change in the internal energy of the system, plus the work that the system has done on its surroundings.
Endergonic	Endergonic means absorbing energy in the form of work. Its etymology stems from the suffix -ergonic, as derived from the Greek root ergon, meaning work, combined with the prefix end-, as derived from the Greek root en, meaning put into. Endergonic reactions are not spontaneous.
Exergonic	Exergonic means to release energy in the form of work. Its etymology stems from the suffix -ergonic, as derived from the Greek root ergon, meaning work, combined with the Greek prefix ex-, meaning out of. By thermodynamic standards, work, a form of energy, is defined as moving from the system to the surroundings (the external region.)
Activation	Activation in (bio-)chemical sciences generally refers to the process whereby something is prepared or excited for a subsequent reaction.

In chemistry, Activation of molecules is required for a chemical reaction to occur. The phrase energy of Activation refers to the energy the reactants must acquire before they can successfully react with each other to produce the products, that is, to reach the transition state.

Methane	Methane is a chemical compound with the molecular formula CH_4. It is the simplest alkane, and the principal component of natural gas. Methane"s bond angles are 109.5 degrees.
Vapor	A Vapor or vapour is a substance in the gas phase at a temperature lower than its critical temperature. This means that the Vapor can be condensed to a liquid or to a solid by increasing its pressure, without reducing the temperature. For example, water has a critical temperature of 374℃ (or 647 K) which is the highest temperature at which liquid water can exist.
Phase diagram	A Phase diagram in physical chemistry, engineering, mineralogy, and materials science is a type of chart used to show conditions at which thermodynamically-distinct phases can occur at equilibrium. In mathematics and physics, "Phase diagram" is used with a different meaning: a synonym for a phase space. Common components of a Phase diagram are lines of equilibrium or phase boundaries, which refer to lines that mark conditions under which multiple phases can coexist at equilibrium.
Triple point	In thermodynamics, the Triple point of a substance is the temperature and pressure at which three phases (for example, gas, liquid, and solid) of that substance coexist in thermodynamic equilibrium. For example, the Triple point of mercury occurs at a temperature of –38.8344 ℃ and a pressure of 0.2 mPa. In addition to the Triple point between solid, liquid, and gas, there can be Triple point s involving more than one solid phase, for substances with multiple polymorphs.
Sublimation	Sublimation of an element or compound is a transition from the solid to gas phase with no intermediate liquid stage. Sublimation is an endothermic phase transition that occurs at temperatures and pressures below the triple point At normal pressures, most chemical compounds and elements possess three different states at different temperatures.
Hemoglobin	Hemoglobin is the iron-containing oxygen-transport metalloprotein in the red blood cells of vertebrates, and the tissues of some invertebrates. In mammals, the protein makes up about 97% of the red blood cell"s dry content, and around 35% of the total content . Hemoglobin transports oxygen from the lungs or gills to the rest of the body where it releases the oxygen for cell use.

Solution	In chemistry, a Solution is a homogeneous mixture composed of two or more substances. In such a mixture, a solute is dissolved in another substance, known as a solvent. Gases may dissolve in liquids, for example, carbon dioxide or oxygen in water.
Colligative properties	Colligative properties are properties of solutions that depend on the number of molecules in a given volume of solvent and not on the properties (e.g. size or mass) of the molecules. Colligative properties include: lowering of vapor pressure; elevation of boiling point; depression of freezing point and osmotic pressure. Measurements of these properties for a dilute aqueous solution of a non-ionized solute such as urea or glucose can lead to accurate determinations of relative molecular masses.
Mole fraction	In chemistry, Mole fraction x (also, and more correctly, known as the amount fraction) is a way of expressing the composition of a mixture. The Mole fraction of each component i is defined as its amount of substance n_i divided by the total amount of substance in the system, n $$x_i \overset{\text{def}}{=} \frac{n_i}{n}$$ where $$n = \sum_i n_i$$ The sum is over all components, including the solvent in the case of a chemical solution. As an example, if a mixture is obtained by dissolving 10 moles of sucrose in 90 moles of water, the Mole fraction of sucrose in that mixture is 0.1.
Kinetic theory	Kinetic theory attempts to explain macroscopic properties of gases, such as pressure, temperature by considering their molecular composition and motion. Essentially, the theory posits that pressure is due not to static repulsion between molecules, as was Isaac Newton"s conjecture, but due to collisions between molecules moving at different velocities. Kinetic theory is also known as the Kinetic-Molecular Theory or the Collision Theory or the Kinetic-Molecular Theory of Gases.
Chemical potential	Chemical potential, symbolized by μ, is a quantity first described by the American engineer, chemist and mathematical physicist Josiah Williard Gibbs. He defined it as follows: Gibbs noted also that for the purposes of this definition, any chemical element or combination of elements in given proportions may be considered a substance, whether capable or not of existing by itself as a homogeneous body. Chemical potential is also referred to as partial molar Gibbs energy (.

Chapter 5. Solutions

Thermodynamic	In physics, Thermodynamic s ">power") is the study of the conversion of energy into work and heat and its relation to macroscopic variables such as temperature,volume and pressure. Its underpinnings, based upon statistical predictions of the collective motion of particles from their microscopic behavior, is the field of statistical Thermodynamic s, a branch of statistical mechanics. Historically, Thermodynamic s developed out of need to increase the efficiency of early steam engines.
Activation	Activation in (bio-)chemical sciences generally refers to the process whereby something is prepared or excited for a subsequent reaction.
	In chemistry, Activation of molecules is required for a chemical reaction to occur. The phrase energy of Activation refers to the energy the reactants must acquire before they can successfully react with each other to produce the products, that is, to reach the transition state.
Liquid	Liquid is one of the principal states of matter. A Liquid is a fluid that has the particles loose and can freely form a distinct surface at the boundaries of its bulk material. The surface is a free surface where the Liquid is not constrained by a container.
Ideal solution	In chemistry, an Ideal solution or ideal mixture is a solution in which the enthalpy of solution (or "enthalpy of mixing") is zero; the closer to zero the enthalpy of solution is, the more "ideal" the behavior of the solution becomes. Equivalently, an ideal mixture is one in which the activity coefficients (which measure deviation from ideality) are equal to one.
	The concept of an Ideal solution is fundamental to chemical thermodynamics and its applications, such as the use of colligative properties.
Gas	In physics, a Gas is a state of matter, consisting of a collection of particles (molecules, atoms, ions, electrons, etc.) without a definite shape or volume that are in more or less random motion.
	Due to the electronic nature of the aforementioned particles, a "force field" is present throughout the space around them.
Activity	In chemical thermodynamics Activity is a measure of the "effective concentration" of a species in a mixture. By convention, it is a dimensionless quantity. The Activity of pure substances in condensed phases (solid or liquids) is normally taken as unity.
Activity coefficient	An Activity coefficient is a factor used in thermodynamics to account for deviations from ideal behaviour in a mixture of chemical substances. In an ideal mixture the interactions between each pair of chemical species are the same (or more formally, the enthalpy of mixing is zero) and, as a result, properties of the mixtures can be expressed directly in terms of simple concentrations or partial pressures of the substances present e.g. Raoult"s law. Deviations from ideality are accommodated by modifying the concentration by an Activity coefficient.

Boiling-point elevation	Boiling-point elevation describes the phenomenon that the boiling point of a liquid (a solvent) will be higher when another compound is added, meaning that a solution has a higher boiling point than a pure solvent. This happens whenever a non-volatile solute, such as a salt, is added to a pure solvent, such as water. The boiling point can be measured accurately using an ebullioscope.
Freezing-point depression	Freezing-point depression describes the phenomenon that the freezing point of a liquid (a solvent) is depressed when another compound is added, meaning that a solution has a lower freezing point than a pure solvent. This happens whenever a solute is added to a pure solvent, such as water. The phenomenon may be observed in sea water, which due to its salt content remains liquid at temperatures below 0 ℃, the freezing point of pure water.
Pressure	Pressure is the force per unit area applied in a direction perpendicular to the surface of an object. Gauge Pressure is the Pressure relative to the local atmospheric or ambient Pressure. Pressure is an effect which occurs when a force is applied on a surface.
DNA	Deoxyribonucleic acid (DNA) is a nucleic acid that contains the genetic instructions used in the development and functioning of all known living organisms and some viruses. The main role of DNA molecules is the long-term storage of information. DNA is often compared to a set of blueprints or a recipe, or a code, since it contains the instructions needed to construct other components of cells, such as proteins and RNA molecules.
Electrolyte	An Electrolyte is any substance containing free ions that behaves as an electrically conductive medium. Because they generally consist of ions in solution, Electrolyte s are also known as ionic solutions, but molten Electrolyte s and solid Electrolyte s are also possible. Electrolyte s commonly exist as solutions of acids, bases or salts.
Acid	An Acid is traditionally considered any chemical compound that, when dissolved in water, gives a solution with a hydrogen ion activity greater than in pure water, i.e. a pH less than 7.0. That approximates the modern definition of Johannes Nicolaus Brønsted and Martin Lowry, who independently defined an Acid as a compound which donates a hydrogen ion (H^+) to another compound (called a base.) Common examples include acetic Acid and sulfuric Acid (used in car batteries.)
Heat	In physics and thermodynamics, Heat is the process of energy transfer from one body or system to another due to a difference in temperature. In thermodynamics, the quantity TdS is used as a representative measure of the (inexact) Heat differential δQ, which is the absolute temperature of an object multiplied by the differential quantity of a system"s entropy measured at the boundary of the object. A related term is thermal energy, loosely defined as the energy of a body that increases with its temperature.

Atmosphere	The standard Atmosphere is an international reference pressure defined as 101,325 Pa and formerly used as unit of pressure (symbol: atm.) For practical purposes it has been replaced by the bar which is 100,000 Pa. The difference of about 1% is not significant for many applications, and is within the error range of common pressure gauges.
Ionic atmosphere	Ionic atmosphere is a concept employed in the use of the Debye-Hückel equation which explains the conductivity behaviour of electrolytic solutions.
	If an electrical potential is applied to an electrolytic solution, a positive ion will move towards the negative electrode and drag along an entourage of negative ions with it. The more concentrated the solution, the closer these negative ions are to the positive ion and thus the greater the resistance experienced by the positive ion.
Ionic strength	The Ionic strength of a solution is a measure of the concentration of ions in that solution. Ionic compounds, when dissolved in water, dissociate into ions. The total electrolyte concentration in solution will affect important properties such as the dissociation or the solubility of different salts.
Ion	An Ion is an atom or molecule where the total number of electrons is not equal to the total number of protons, giving it a net positive or negative electrical charge.
	Since protons are positively charged and electrons are negatively charged, if there are more electrons than protons, the atom or molecule will be negatively charged. This is called an an Ion , from the Greek á¼€vÎ¬ , meaning "up".
Passive transport	Passive transport means moving biochemicals and atomic or molecular substances across the cell membrane. Unlike active transport, this process does not involve chemical energy. The four main kinds of Passive transport are diffusion, facilitated diffusion, filtration and osmosis.
Isoelectric point	The Isoelectric point, sometimes abbreviated to IEP, is the pH at which a particular molecule or surface carries no net electrical charge.
	Amphoteric molecules called zwitterions contain both positive and negative charges depending on the functional groups present in the molecule. The net charge on the molecule is affected by pH of their surrounding environment and can become more positively or negatively charged due to the loss or gain of protons (H^+.)
Bilayer	A Bilayer is a double layer of closely packed atoms or molecules. They properties of Bilayer s are studied in condensed matter physics, often in the context of semiconductor devices, where two distinct materials are united to form junctions (such as p-n junctions, Schottky junctions, ...).
Ionization	Ionization is the physical process of converting an atom or molecule into an ion by adding or removing charged particles such as electrons or other ions. This is often confused with dissociation (chemistry.)

The process works slightly differently depending on whether an ion with a positive or a negative electric charge is being produced.

Chapter 6. Chemical Equilibrium

Chemical equilibrium	In a chemical process, Chemical equilibrium is the state in which the chemical activities or concentrations of the reactants and products have no net change over time. Usually, this would be the state that results when the forward chemical process proceeds at the same rate as their reverse reaction. The reaction rates of the forward and reverse reactions are generally not zero but, being equal, there are no net changes in any of the reactant or product concentrations.
Equilibrium constant	Stability constants, formation constants, binding constants, association constants and dissociation constants are all types of Equilibrium constant.
Reaction quotient	In chemistry, Reaction quotient: Q_r is a function of the extent of reaction: ξ, the relative proportion of products and reactants present in the reaction mixture at some instant of time. For a chemical mixture with certain initial concentrations of reactants and products, it is useful to know if the reaction will shift to the right/in the forward direction (increasing the concentrations of the products) or if it will shift to the left/in the reverse direction (increasing the concentrations of the reactants.) Given a general equilibrium expression such as kA + mB ...
Fugacity	Fugacity is a measure of a chemical potential in the form of "adjusted pressure." It reflects the tendency of a substance to prefer one phase (liquid, solid and can be literally defined as "the tendency to flee or escape". At a fixed temperature and pressure, a homogeneous substance will have a different Fugacity for each phase. The phase with the lowest Fugacity will be the most favorable, and will have the lowest Gibbs free energy.
Thermodynamic	In physics, Thermodynamic s ">power") is the study of the conversion of energy into work and heat and its relation to macroscopic variables such as temperature,volume and pressure. Its underpinnings, based upon statistical predictions of the collective motion of particles from their microscopic behavior, is the field of statistical Thermodynamic s, a branch of statistical mechanics. Historically, Thermodynamic s developed out of need to increase the efficiency of early steam engines.
Thermodynamic equilibrium	In thermodynamics, a thermodynamic system is said to be in Thermodynamic equilibrium when it is in thermal equilibrium, mechanical equilibrium, and chemical equilibrium. The local state of a system at Thermodynamic equilibrium is determined by the values of its intensive parameters, as pressure, temperature, etc. Specifically, Thermodynamic equilibrium is characterized by the minimum of a thermodynamic potential, such as the Helmholtz free energy, i.e. systems at constant temperature and volume: A = U - TS. Or as the Gibbs free energy, i.e. systems at constant pressure and temperature: G = H - TS.

The process that leads to a Thermodynamic equilibrium is called thermalization.

Pressure	Pressure is the force per unit area applied in a direction perpendicular to the surface of an object. Gauge Pressure is the Pressure relative to the local atmospheric or ambient Pressure. Pressure is an effect which occurs when a force is applied on a surface.
Catalyst	Catalysis is the process in which the rate of a chemical reaction is either increased or decreased by means of a chemical substance known as a Catalyst. Unlike other reagents that participate in the chemical reaction, a Catalyst is not consumed by the reaction itself. The Catalyst may participate in multiple chemical transformations.
Chemical kinetics	Chemical kinetics is the study of rates of chemical processes. Chemical kinetics includes investigations of how different experimental conditions can influence the speed of a chemical reaction and yield information about the reaction"s mechanism and transition states, as well as the construction of mathematical models that can describe the characteristics of a chemical reaction. In 1864, Peter Waage and Cato Guldberg pioneered the development of Chemical kinetics by formulating the law of mass action, which states that the speed of a chemical reaction is proportional to the quantity of the reacting substances.
Dissociation	Dissociation in chemistry and biochemistry is a general process in which ionic compounds (complexes, molecules ions usually in a reversible manner. When a Bronsted-Lowry acid is put in water, a covalent bond between an electronegative atom and a hydrogen atom is broken by heterolytic fission, which gives a proton and a negative ion. Dissociation is the opposite of association and recombination.
Dissociation constant	In chemistry and biochemistry, a Dissociation constant is a specific type of equilibrium constant that measures the propensity of a larger object to separate (dissociate) reversibly into smaller components, as when a complex falls apart into its component molecules, or when a salt splits up into its component ions. The Dissociation constant is usually denoted K_d and is the inverse of the association constant. In the special case of salts, the Dissociation constant can also be called an ionization constant.
Steady state	In chemistry, a Steady state is a situation in which all state variables are constant in spite of ongoing processes that strive to change them. For an entire system to be at Steady state, i.e. for all state variables of a system to be constant, there must be a flow through the system (compare mass balance.) One of the most simple examples of such a system is the case of a bathtub with the tap open but without the bottom plug: after a certain time the water flows in and out at the same rate, so the water level (the state variable being Volume) stabilizes and the system is at Steady state.

Eutectic point

The melting point of a mixture of two or more solids (such as an alloy) depends on the relative proportions of its ingredients. A eutectic or eutectic mixture is a mixture at such proportions that the melting point is a local temperature minimum, which means that all the constituents crystallize simultaneously at this temperature from molten liquid solution. Such a simultaneous crystallization of a eutectic mixture is known as a eutectic reaction, the temperature at which it takes place is the eutectic temperature, and the composition and temperature at which it takes place is called the Eutectic point.

Electrochemical cell	An Electrochemical cell is a device used for generating an electromotive force (voltage) and current from chemical reactions inducing a chemical reaction by a flow of current. The current is caused by the reactions releasing and accepting electrons at the different ends of a conductor. A common example of an Electrochemical cell is a standard 1.5-volt battery.
Electrochemistry	Electrochemistry is a branch of chemistry that studies chemical reactions which take place in a solution at the interface of an electron conductor (a metal or a semiconductor) and an ionic conductor (the electrolyte), and which involve electron transfer between the electrode and the electrolyte or species in solution.
	If a chemical reaction is driven by an external applied voltage, as in electrolysis, or if a voltage is created by a chemical reaction as in a battery, it is an electrochemical reaction. Chemical reactions where electrons are transferred between molecules are called oxidation/reduction (redox) reactions.
Anode	An Anode is an electrode through which electric current flows into a polarized electrical device. Mnemonic: ACID (Anode Current Into Device.) Electrons flow in the opposite direction to the electric current (flow of hypothetical positive charge)
	A widespread misconception is that Anode polarity is always positive.
Cathode	A Cathode is an electrode through which electric current flows out of a polarized electrical device. Mnemonic: CCD (Cathode Current Departs.)
	A widespread misconception is that Cathode polarity is always negative.
Daniell cell	The Daniell cell, also called the gravity cell or crowfoot cell was invented in 1836 by John Frederic Daniell, who was a British chemist and meteorologist.
Electrode	An Electrode is an electrical conductor used to make contact with a nonmetallic part of a circuit (e.g. a semiconductor, an electrolyte or a vacuum.) The word was coined by the scientist Michael Faraday from the Greek words elektron and hodos, a way.
	An Electrode in an electrochemical cell is referred to as either an anode or a cathode, words that were also coined by Faraday.
Electrode potential	Electrode potential, E, in electrochemistry, according to an IUPAC definition, is the electromotive force of a cell built of two electrodes:
	• on the left-hand side is the standard hydrogen electrode, and • on the right-hand side is the electrode the potential of which is being defined.
	By convention:
	$$E_{Cell} := E_{Right} - E_{Left}$$

From the above, for the cell with the standard hydrogen electrode (potential of 0 by convention), one obtains:

$$E_{Cell} = E_{Right} - 0 = E_{Electrode}$$

Electrode potential is measured in volt (V.)

Three-electrode setup for measurement of Electrode potential

The measurement is generally conducted using a three-electrode setup :

1. Working electrode
2. Counter electrode
3. Reference electrode (standard hydrogen electrode or an equivalent)

The measured potential of the working electrode may be either that at equilibrium on the working electrode ("reversible potential"), or a potential with a non-zero net reaction on the working electrode but zero net current ("corrosion potential", "mixed potential"), or a potential with a non-zero net current on the working electrode.

Galvanic cell	The Galvanic cell is a part of a battery consisting of an electrochemical cell with two different metals connected by a salt bridge or a porous disk between the individual half-cells. It is sometimes also called a Voltaic cell.
	Common usage of the word battery has evolved to include a single Galvanic cell but the first batteries had many Galvanic cell s.
Salt bridge	A Salt bridge in chemistry, is a laboratory device used to connect the oxidation and reduction half-cells of a galvanic cell (voltaic cell), a type of electrochemical cell. Salt bridge s usually come in two types: glass tube and filter paper.
	One type of Salt bridge consists of a U-shaped glass tube filled with a relatively inert electrolyte, usually potassium chloride or sodium chloride is used, although the diagram here illustrates the use of a potassium nitrate solution.
Standard hydrogen electrode	The Standard hydrogen electrode is a redox electrode which forms the basis of the thermodynamic scale of oxidation-reduction potentials. Its absolute electrode potential is estimated to be 4.44 ± 0.02 V at 25 °C, but to form a basis for comparison with all other electrode reactions, hydrogen"s standard electrode potential is declared to be zero at all temperatures. Potentials of any other electrodes are compared with that of the Standard hydrogen electrode at the same temperature.
Gas	In physics, a Gas is a state of matter, consisting of a collection of particles (molecules, atoms, ions, electrons, etc.) without a definite shape or volume that are in more or less random motion.

Due to the electronic nature of the aforementioned particles, a "force field" is present throughout the space around them.

Reduction potential	Reduction potential is the tendency of a chemical species to acquire electrons and thereby be reduced. Each species has its own intrinsic Reduction potential; the more positive the potential, the greater the species" affinity for electrons and tendency to be reduced. In aqueous solutions, the Reduction potential is the tendency of the solution to either gain or lose electrons when it is subject to change by introduction of a new species.
Thermodynamic	In physics, Thermodynamic s ">power") is the study of the conversion of energy into work and heat and its relation to macroscopic variables such as temperature,volume and pressure. Its underpinnings, based upon statistical predictions of the collective motion of particles from their microscopic behavior, is the field of statistical Thermodynamic s, a branch of statistical mechanics. Historically, Thermodynamic s developed out of need to increase the efficiency of early steam engines.
Activation	Activation in (bio-)chemical sciences generally refers to the process whereby something is prepared or excited for a subsequent reaction. In chemistry, Activation of molecules is required for a chemical reaction to occur. The phrase energy of Activation refers to the energy the reactants must acquire before they can successfully react with each other to produce the products, that is, to reach the transition state.
Potential temperature	The Potential temperature of a parcel of fluid at pressure P is the temperature that the parcel would acquire if adiabatically brought to a standard reference pressure P_0, usually 1000 millibars. The Potential temperature is denoted θ and, for air, is often given by $$\theta = T \left(\frac{P_0}{P} \right)^{\frac{R}{c_p}},$$ where T is the current absolute temperature (in K) of the parcel, R is the gas constant of air, and c_p is the specific heat capacity at a constant pressure. The concept of Potential temperature applies to any stratified fluid.
Fuel cell	A Fuel cell is an electrochemical conversion device. It produces electricity from fuel (on the anode side) and an oxidant (on the cathode side), which react in the presence of an electrolyte. The reactants flow into the cell, and the reaction products flow out of it, while the electrolyte remains within it.

Electrocatalyst	An Electrocatalyst is a catalyst that participates in electrochemical reaction. Catalyst materials modify and increase the rate of chemical reactions without being consumed in the process. Electrocatalyst are a specific form of catalysts that function at electrode surfaces or may be the electrode surface itself.
Acid	An Acid is traditionally considered any chemical compound that, when dissolved in water, gives a solution with a hydrogen ion activity greater than in pure water, i.e. a pH less than 7.0. That approximates the modern definition of Johannes Nicolaus Brønsted and Martin Lowry, who independently defined an Acid as a compound which donates a hydrogen ion (H^+) to another compound (called a base.) Common examples include acetic Acid and sulfuric Acid (used in car batteries.)
Enzymes	Enzymes are biomolecules that catalyze (i.e., increase the rates of) chemical reactions. Nearly all known Enzymes are proteins. However, certain RNA molecules can be effective biocatalysts too.
Membrane potential	Membrane potential is the voltage difference (or electrical potential difference) between the interior and exterior of a cell. Because the fluid inside and outside a cell is highly conductive, while a cell"s plasma membrane is highly resistive, the voltage change in moving from a point outside to a point inside occurs largely within the narrow width of the membrane itself. Therefore, it is common to speak of the Membrane potential as the voltage across the membrane.
Faraday constant	In physics and chemistry, the Faraday constant is the magnitude of electric charge per mole of electrons. While most uses of the Faraday constant, denoted F, have been replaced by the standard SI unit, the coulomb, the Faraday is still widely used in calculations in electrochemistry. It has the currently accepted value: $$F = 96\,485.3399(24)\,\frac{C}{mol}$$ The constant F has a simple relation to two other physical constants: $$F = N_A e$$ where $$N_A = 6.022 \times 10^{23}\,mol^{-1}$$ $$e = 1.602 \times 10^{-19}\,C$$ N_A is the Avogadro constant, and e is the elementary charge or the magnitude of the charge of an electron.

Enthalpy

In thermodynamics and molecular chemistry, the Enthalpy is a thermodynamic property of a fluid. It can be used to calculate the heat transfer during a quasistatic process taking place in a closed thermodynamic system under constant pressure. Enthalpy H is an arbitrary concept but the Enthalpy change ΔH is more useful because it is equal to the change in the internal energy of the system, plus the work that the system has done on its surroundings.

Entropy

Entropy is a concept applied across physics, information theory, mathematics and other branches of science and engineering. The following definition is shared across all these fields:

$$S = -k \sum_i P_i \ln P_i$$

where S is the conventional symbol for Entropy. The sum runs over all microstates consistent with the given macrostate and P_i is the probability of the ith microstate.

Acid	An Acid is traditionally considered any chemical compound that, when dissolved in water, gives a solution with a hydrogen ion activity greater than in pure water, i.e. a pH less than 7.0. That approximates the modern definition of Johannes Nicolaus Brønsted and Martin Lowry, who independently defined an Acid as a compound which donates a hydrogen ion (H^+) to another compound (called a base.) Common examples include acetic Acid and sulfuric Acid (used in car batteries.)
Base	In chemistry, a Base is most commonly thought of as an aqueous substance that can accept hydrogen ions..Bases are also the oxides or hydroxides of metals.A soluble Base is also often referred to as an alkali if OH^- ions are involved. This refers to the Brønsted-Lowry theory of acids and bases. Alternate definitions of bases include electron pair donors (Lewis), as sources of hydroxide anions (Arrhenius.)
Hydronium	In chemistry, Hydronium is the common name for the aqueous cation H_3O^+, the simplest type of oxonium ion, produced by protonation of water. It is the positive ion present when an Arrhenius acid is dissolved in water, as Arrhenius acid molecules in solution give up a proton (a positive hydrogen ion, H^+) to the surrounding water molecules (H_2O.)
	It is the presence of Hydronium ion relative to hydroxide that determines a solution"s pH. The molecules in pure water auto-dissociate into Hydronium and hydroxide ions in the following equillibrium:
	$$2H_2O \; OH^- + H_3O^+$$
	In pure water, there is an equal number of hydroxide and Hydronium ions, and the pH is perfectly neutral (7.0.)
Lewis acid	A Lewis acid is a chemical compound, A, that can accept a pair of electrons from a Lewis base, B, that acts as an electron-pair donor, forming an adduct, AB.
	$$A + :B \rightarrow A\text{--}B$$
	Gilbert N. Lewis proposed this definition, which is based on chemical bonding theory, in 1923. Brønsted-Lowry acid-base theory was published in the same year. The two theories are distinct but complementary to each other as a Lewis base is also a Brønsted-Lowry base, but a Lewis acid need not be a Brønsted-Lowry acid.
Ion	An Ion is an atom or molecule where the total number of electrons is not equal to the total number of protons, giving it a net positive or negative electrical charge.

Since protons are positively charged and electrons are negatively charged, if there are more electrons than protons, the atom or molecule will be negatively charged. This is called an an Ion , from the Greek á¼€vÎ¬ , meaning "up".

| Dissociation | Dissociation in chemistry and biochemistry is a general process in which ionic compounds (complexes, molecules ions usually in a reversible manner. When a Bronsted-Lowry acid is put in water, a covalent bond between an electronegative atom and a hydrogen atom is broken by heterolytic fission, which gives a proton and a negative ion. Dissociation is the opposite of association and recombination. |

| Dissociation constant | In chemistry and biochemistry, a Dissociation constant is a specific type of equilibrium constant that measures the propensity of a larger object to separate (dissociate) reversibly into smaller components, as when a complex falls apart into its component molecules, or when a salt splits up into its component ions. The Dissociation constant is usually denoted K_d and is the inverse of the association constant. In the special case of salts, the Dissociation constant can also be called an ionization constant. |

| Relaxation | In nuclear magnetic resonance (NMR) spectroscopy and magnetic resonance imaging (MRI) the term Relaxation describes several processes by which nuclear magnetization prepared in a non-equilibrium state return to the equilibrium distribution. In other words, Relaxation describes how fast spins "forget" the direction in which they are oriented. The rates of this spin Relaxation can be measured in both spectroscopy and imaging applications. |

| Chemical kinetics | Chemical kinetics is the study of rates of chemical processes. Chemical kinetics includes investigations of how different experimental conditions can influence the speed of a chemical reaction and yield information about the reaction"s mechanism and transition states, as well as the construction of mathematical models that can describe the characteristics of a chemical reaction. In 1864, Peter Waage and Cato Guldberg pioneered the development of Chemical kinetics by formulating the law of mass action, which states that the speed of a chemical reaction is proportional to the quantity of the reacting substances. |

| Hydrolysis | Hydrolysis is a chemical reaction during which one or more water molecules are split into hydrogen and hydroxide ions, which may go on to participate in further reactions. It is the type of reaction that is used to break down certain polymers, especially those made by step-growth polymerization. Such polymer degradation is usually catalysed by either acid e.g. concentrated sulfuric acid (H_2SO_4) or alkali e.g. sodium hydroxide (NaOH) attack, often increasing with their strength or pH.

Hydrolysis is distinct from hydration, where the hydrated molecule does not "lyse" (break into two new compounds.) |

Carbonic acid	Carbonic acid (ancient name acid of air or aerial acid) has the formula H_2CO_3. It is also a name sometimes given to solutions of carbon dioxide in water, which contain small amounts of H_2CO_3. The salts of Carbonic acid s are called bicarbonates (or hydrogen carbonates) and carbonates.
Diprotic acid	A Diprotic acid is an acid such as H_2SO_4 (sulfuric acid) that happens to contain within its molecular structure two hydrogen atoms capable of dissociating (i.e. ionizable) in water. The complete dissociation of Diprotic acid s is of the same form as sulfuric acid:

$$H_2SO_4 \rightarrow H^+(aq) + HSO_4^-(aq) \quad K_a = 1 \times 10^3$$

$$HSO_4^- \rightarrow H^+(aq) + SO_4^{2-}(aq) \quad K_a = 1 \times 10^{-2}$$

	The dissociation does not happen all at once due to the two stages of dissociation having different K_a values. The first dissociation will, in the case of sulfuric acid, occur completely, but the second one will not.
Buffer solution	A Buffer solution is an aqueous solution consisting of a mixture of a weak acid and its conjugate base or a weak base and its conjugate acid. It has the property that the pH of the solution changes very little when a small amount of acid or base is added to it. Buffer solution s are used as a means of keeping pH at a nearly constant value in a wide variety of chemical applications.
Isoelectric point	The Isoelectric point, sometimes abbreviated to IEP, is the pH at which a particular molecule or surface carries no net electrical charge.
	Amphoteric molecules called zwitterions contain both positive and negative charges depending on the functional groups present in the molecule. The net charge on the molecule is affected by pH of their surrounding environment and can become more positively or negatively charged due to the loss or gain of protons (H^+.)
Component	In thermodynamics, a Component is a chemically distinct constituent of a system. Calculating the number of components in a system is necessary, for example, when applying Gibbs phase rule in determination of the number of degrees of freedom of a system.
	The number of components is equal to the number of independent chemical constituents, minus the number of chemical reactions between them, minus the number of any constraints (like charge neutrality or balance of molar quantities.)
Enzymes	Enzymes are biomolecules that catalyze (i.e., increase the rates of) chemical reactions. Nearly all known Enzymes are proteins. However, certain RNA molecules can be effective biocatalysts too.
Hemoglobin	Hemoglobin is the iron-containing oxygen-transport metalloprotein in the red blood cells of vertebrates, and the tissues of some invertebrates.

In mammals, the protein makes up about 97% of the red blood cell"s dry content, and around 35% of the total content . Hemoglobin transports oxygen from the lungs or gills to the rest of the body where it releases the oxygen for cell use.

Equilibrium constant	Stability constants, formation constants, binding constants, association constants and dissociation constants are all types of Equilibrium constant.

Order	Order in a crystal lattice is the arrangement of some property with respect to atomic positions. It arises in charge ordering, spin ordering, magnetic ordering, and compositional ordering. It is a thermodynamic entropy concept often displayed by a second Order phase transition.
Critical opalescence	Critical opalescence is a phenomenon which arises in the region of a continuous phase transition. Originally reported by Thomas Andrews in 1869 for the liquid-gas transition in carbon dioxide, many other examples have been discovered since. The phenomenon is most commonly demonstrated in binary fluid mixtures, such as methanol and cyclohexane.
DNA	Deoxyribonucleic acid (DNA) is a nucleic acid that contains the genetic instructions used in the development and functioning of all known living organisms and some viruses. The main role of DNA molecules is the long-term storage of information. DNA is often compared to a set of blueprints or a recipe, or a code, since it contains the instructions needed to construct other components of cells, such as proteins and RNA molecules.
Molecularity	Molecularity in chemistry is the number of colliding molecular entities that are involved in a single reaction step. While the order of a reaction is derived experimentally, the Molecularity is a theoretical concept and can only be applied to elementary reactions. In elementary reactions, the reaction order, the Molecularity and the stoichiometric coefficient are the same, although only numerically, because they are different concepts.
Reaction mechanism	In chemistry, a Reaction mechanism is the step by step sequence of elementary reactions by which overall chemical change occurs . Although only the net chemical change is directly observable for most chemical reactions, experiments can often be designed that suggest the possible sequence of steps in a Reaction mechanism. A mechanism describes in detail exactly what takes place at each stage of a chemical transformation.
Rate-determining step	The Rate-determining step is a chemistry term for the slowest step in a chemical reaction. The Rate-determining step is often compared to the neck of a funnel; the rate at which water flows through the funnel is determined by the width of the neck, not by the speed at which water is poured in. Similarly, the rate of reaction depends on the rate of the slowest step.
Elementary Reaction	An Elementary reaction is a chemical reaction in which one or more of the chemical species react directly to form products in a single reaction step and with a single transition state. In a unimolecular Elementary reaction a molecule, A, dissociates or isomerises to form the products(s.) $$A \rightarrow \text{products}$$

The rate of such a reaction, at constant temperature, is proportional to the concentration of the species A

$$\frac{d[A]}{dt} = -k[A]$$

In a bimolecular Elementary reaction, two atoms, molecules, ions or radicals, A and B, react together to form the product(s)

$$A + B \rightarrow \text{products}$$

The rate of such a reaction, at constant temperature, is proportional to the product of the concentrations of the species A and B.

$$\frac{d[A]}{dt} = \frac{d[B]}{dt} = -k[A][B]$$

This rate expression can be derived from first principles by using collision theory.

Reversible reaction

A Reversible reaction is a chemical reaction that results in an equilibrium mixture of reactants and products. For a reaction involving two reactants and two products this can be expressed symbolically as

$$aA + bB \rightleftharpoons cC + dD$$

A and B can react to form C and D or, in the reverse reaction, C and D can react to form A and B. This is distinct from reversible process in thermodynamics.

The concentrations of reactants and products in an equilibrium mixture are determined by the analytical concentrations of the reagents (A and B or C and D) and the equilibrium constant, K. The magnitude of the equilibrium constant depends on the Gibbs free energy change for the reaction.

Microscopic reversibility

The principle of Microscopic reversibility in chemistry states that in a reversible reaction the mechanism in one direction is exactly the reverse of the mechanism in the other direction. A result of Microscopic reversibility is that the series of transition states and intermediates of the forward reaction are mirrored in reverse order in the reverse reaction. The principle does not apply to reactions where photochemical excitation is the first step.

Chain reaction	A Chain reaction is a sequence of reactions where a reactive product or by-product causes additional reactions to take place. In a Chain reaction positive feedback leads to a self-amplifying chain of events. Examples of Chain reaction s include:
	• The neutron-fission Chain reaction a neutron plus a fissionable atom causes a fission resulting in a larger number of neutrons than was consumed in the initial reaction. This reaction could continue if the number of neutron produced in a single reaction is capable of producin another fission. If not the reaction will stop. If the number of fissions produced is more than one then the reaction is said to be supercritical and the number of fissions would increase exponentially. This is the principle for an atomic bomb.
	• For example in a chemical reaction every step of $H_2 + Cl_2$ Chain reaction consumes one molecule of H_2 or Cl_2, one free radical HÂ· or ClÂ· producing one HCl molecule and another free radical.
	• Electron avalanche process: Collisions of free electrons in a strong electric field releasing "new" electrons to undergo the same process in successive cycles.
	• A cascading failure, a failure in a system of interconnected parts, for example a power transmission grid, where the service provided depends on the operation of a preceding part, and the failure of a preceding part can trigger the failure of successive parts.
	• Polymerase Chain reaction a technique used in molecular biology to amplify (make many copies of) a piece of DNA by in vitro enzymatic replication using a DNA polymerase.
	.
Arrhenius equation	The Arrhenius equation is a simple, but remarkably accurate, formula for the temperature dependence of the rate constant, and therefore, rate of a chemical reaction. The equation was first proposed by the Dutch chemist J. H. van "t Hoff in 1884; five years later in 1889, the Swedish chemist Svante Arrhenius provided a physical justification and interpretation for it. Nowadays it is best seen as an empirical relationship.
Activation	Activation in (bio-)chemical sciences generally refers to the process whereby something is prepared or excited for a subsequent reaction.
	In chemistry, Activation of molecules is required for a chemical reaction to occur. The phrase energy of Activation refers to the energy the reactants must acquire before they can successfully react with each other to produce the products, that is, to reach the transition state.
Activation energy	In chemistry, Activation energy is a term introduced in 1889 by the Swedish scientist Svante Arrhenius, that is defined as the energy that must be overcome in order for a chemical reaction to occur. Arrhenius" research was a follow up of the theories of reaction rate by Serbian physicist Nebojsa Lekovic. Activation energy may also be defined as the minimum energy required to start a chemical reaction.
Activated complex	In chemistry an Activated complex is a transitional structure in a chemical reaction that results from an effective transfusion between molecules and that persists while unaccounted for bonds are breaking and new bonds are forming. It is therefore a range of molecular geometries along the reaction coordinate.

Potential energy surface	A Potential energy surface is generally used within the adiabatic or Born-Oppenheimer approximation in quantum mechanics and statistical mechanics to model chemical reactions and interactions in simple chemical and physical systems. The "(hyper)surface" name comes from the fact that the total energy of an atom arrangement can be represented as a curve or (multidimensional) surface, with atomic positions as variables. The best visualization for a layman would be to think of a landscape, where going North-South and East-West are two independent variables (the equivalent of two geometrical parameters of the molecule), and the height of the land we are on would be the energy associated with a given value of such variables.
Preexponential factor	In chemical kinetics, the preexponential factor or A factor is the pre-exponential constant in the Arrhenius equation, an empirical relationship between temperature and rate coefficient. It is usually designated by A when determined from experiment, while Z is usually left for collision frequency.
	For a first order reaction it has units of s^{-1}, for that reason it is often called frequency factor.
Reaction coordinate	In chemistry, a Reaction coordinate is an abstract one-dimensional coordinate which represents progress along a reaction pathway. It is usually a geometric parameter that changes during the conversion of one or more molecular entities.
	These coordinates can sometimes represent a real coordinate system (such as bond length, bond angle...), although, for more complex reactions especially, this can be difficult (and non geometric parameters are used, e.g., bond order.)
Collision theory	The Collision theory, proposed by Max Trautz and William Lewis in 1916 and 1918, qualitatively explains how chemical reactions occur and why reaction rates differ for different reactions. This theory is based on the idea that reactant particles must collide for a reaction to occur, but only a certain fraction of the total collisions have the energy to connect effectively and cause the reactants to transform into products. This is because only a portion of the molecules have enough energy and the right orientation (or "angle") at the moment of impact to break any existing bonds and form new ones.
Acid	An Acid is traditionally considered any chemical compound that, when dissolved in water, gives a solution with a hydrogen ion activity greater than in pure water, i.e. a pH less than 7.0. That approximates the modern definition of Johannes Nicolaus Brønsted and Martin Lowry, who independently defined an Acid as a compound which donates a hydrogen ion (H^+) to another compound (called a base.) Common examples include acetic Acid and sulfuric Acid (used in car batteries.)

Base	In chemistry, a Base is most commonly thought of as an aqueous substance that can accept hydrogen ions..Bases are also the oxides or hydroxides of metals.A soluble Base is also often referred to as an alkali if OH^- ions are involved. This refers to the Brønsted-Lowry theory of acids and bases. Alternate definitions of bases include electron pair donors (Lewis), as sources of hydroxide anions (Arrhenius.)
Steric factor	Steric factor, P is a term used in collision theory. It is defined as the ratio between the experimental value of the rate constant and the one predicted by collision theory. It can also be defined as the ratio between the preexponential factor and the collision frequency, and it is most often less than unity.
Enthalpy	In thermodynamics and molecular chemistry, the Enthalpy is a thermodynamic property of a fluid. It can be used to calculate the heat transfer during a quasistatic process taking place in a closed thermodynamic system under constant pressure. Enthalpy H is an arbitrary concept but the Enthalpy change ΔH is more useful because it is equal to the change in the internal energy of the system, plus the work that the system has done on its surroundings.
Transition state	The Transition state of a chemical reaction is a particular configuration along the reaction coordinate. It is defined as the state corresponding to the highest energy along this reaction coordinate. At this point, assuming a perfectly irreversible reaction, colliding reactant molecules will always go on to form products .
Thermodynamic	In physics, Thermodynamic s ">power") is the study of the conversion of energy into work and heat and its relation to macroscopic variables such as temperature,volume and pressure. Its underpinnings, based upon statistical predictions of the collective motion of particles from their microscopic behavior, is the field of statistical Thermodynamic s, a branch of statistical mechanics. Historically, Thermodynamic s developed out of need to increase the efficiency of early steam engines.
Kinetic isotope effect	The Kinetic isotope effect is a dependence of the rate of a chemical reaction on the isotopic identity of an atom in a reactant. It is also called "isotope fractionation," although this term is somewhat broader in meaning. A Kinetic isotope effect involving hydrogen and deuterium is represented as:$$KIE = \frac{k_H}{k_D}$$with k_H and k_D reaction rate constants.
Enzymes	Enzymes are biomolecules that catalyze (i.e., increase the rates of) chemical reactions. Nearly all known Enzymes are proteins. However, certain RNA molecules can be effective biocatalysts too.

Relaxation	In nuclear magnetic resonance (NMR) spectroscopy and magnetic resonance imaging (MRI) the term Relaxation describes several processes by which nuclear magnetization prepared in a non-equilibrium state return to the equilibrium distribution. In other words, Relaxation describes how fast spins "forget" the direction in which they are oriented. The rates of this spin Relaxation can be measured in both spectroscopy and imaging applications.
Temperature jump	A Temperature jump is a piece of apparatus useful in the study of chemical kinetics. It involves the discharging of a capacitor (in the kV range) through a small volume (<ml) solution containing the molecule/reaction to be studied. This causes the solution to rise in temperature by a few degrees in microseconds.
Chemical kinetics	Chemical kinetics is the study of rates of chemical processes. Chemical kinetics includes investigations of how different experimental conditions can influence the speed of a chemical reaction and yield information about the reaction"s mechanism and transition states, as well as the construction of mathematical models that can describe the characteristics of a chemical reaction. In 1864, Peter Waage and Cato Guldberg pioneered the development of Chemical kinetics by formulating the law of mass action, which states that the speed of a chemical reaction is proportional to the quantity of the reacting substances.
Faraday constant	In physics and chemistry, the Faraday constant is the magnitude of electric charge per mole of electrons. While most uses of the Faraday constant, denoted F, have been replaced by the standard SI unit, the coulomb, the Faraday is still widely used in calculations in electrochemistry. It has the currently accepted value: $$F = 96\,485.3399(24)\,\frac{C}{mol}$$ The constant F has a simple relation to two other physical constants: $$F = N_A e$$ where $$N_A = 6.022 \times 10^{23}\,mol^{-1}$$ $$e = 1.602 \times 10^{-19}\,C$$ N_A is the Avogadro constant, and e is the elementary charge or the magnitude of the charge of an electron.

Catalysis	Catalysis is the process in which the rate of a chemical reaction is either increased or decreased by means of a chemical substance known as a catalyst. Unlike other reagents that participate in the chemical reaction, a catalyst is not consumed by the reaction itself. The catalyst may participate in multiple chemical transformations.
Enzymes	Enzymes are biomolecules that catalyze (i.e., increase the rates of) chemical reactions. Nearly all known Enzymes are proteins. However, certain RNA molecules can be effective biocatalysts too.
Enzyme kinetics	Enzyme kinetics is the study of the chemical reactions that are catalysed by enzymes, with a focus on their reaction rates. The study of an enzyme"s kinetics reveals the catalytic mechanism of this enzyme, its role in metabolism, how its activity is controlled, and how a drug or a poison might inhibit the enzyme.

Enzymes are usually protein molecules that manipulate other molecules -- the enzymes" substrates. |
| Active site | The Active site of an enzyme contains the catalytic and binding sites. The structure and chemical properties of the Active site allow the recognition and binding of the substrate.

The Active site is usually a big pocket or cleft surrounded by amino acid- and other side chains at the surface of the enzyme that contains residues responsible for the substrate specificity (charge, hydrophobicity, steric hindrance) and catalytic residues which often act as proton donors or acceptors or are responsible for binding a cofactor such as PLP, TPP or NAD. The Active site is also the site of inhibition of enzymes |
| Ribozyme | A Ribozyme is an RNA molecule that catalyzes a chemical reaction. Many natural Ribozyme s catalyze either the hydrolysis of one of their own phosphodiester bonds, or the hydrolysis of bonds in other RNAs, but they have also been found to catalyze the aminotransferase activity of the ribosome.

Investigators studying the origin of life have produced Ribozyme s in the laboratory that are capable of catalyzing their own synthesis under very specific conditions, such as an RNA polymerase Ribozyme |
Substrate	In biochemistry, a substrate is a molecule upon which an enzyme acts. Enzymes catalyze chemical reactions involving the substrate(s.) In the case of a single substrate, the substrate binds with the enzyme active site, and an enzyme-substrate complex is formed.
Carbonic acid	Carbonic acid (ancient name acid of air or aerial acid) has the formula H_2CO_3. It is also a name sometimes given to solutions of carbon dioxide in water, which contain small amounts of H_2CO_3. The salts of Carbonic acid s are called bicarbonates (or hydrogen carbonates) and carbonates.
Turnover number	Turnover number has two related meanings:

In enzymology, Turnover number is defined as the maximum number of molecules of substrate that an enzyme can convert to product per catalytic site per unit of time and can be calculated as follows: $k_{cat} = V_{max}/[E]_T$ For example, carbonic anhydrase has a Turnover number of 400,000 to 600,000 s^{-1}, which means that each carbonic anhydrase molecule can produce up to 600,000 molecules of product (CO_2) per second.

In more chemical fields, such as organometallic catalysis, Turnover number is used with a slightly different meaning: the number of moles of substrate that a mole of catalyst can convert before becoming inactivated.

Activity	In chemical thermodynamics Activity is a measure of the "effective concentration" of a species in a mixture. By convention, it is a dimensionless quantity. The Activity of pure substances in condensed phases (solid or liquids) is normally taken as unity.
Catalase	Catalase is a common enzyme found in nearly all living organisms which are exposed to oxygen, where it functions to catalyze the decomposition of hydrogen peroxide to water and oxygen. Catalase has one of the highest turnover numbers of all enzymes; one molecule of Catalase can convert millions of molecules of hydrogen peroxide to water and oxygen per second.
	Catalase is a tetramer of four polypeptide chains, each over 500 amino acids long.
Chymotrypsin	Chymotrypsin is a digestive enzyme that can perform proteolysis. Chymotrypsin cleaves peptides at the carboxyl side of tyrosine, tryptophan, and phenylalanine because these three amino acids contain aromatic rings, which fit into a "hydrophobic pocket" in the enzyme. Over time, Chymotrypsin also hydrolyzes other amide bonds, particularly those with leucine-donated carboxyls.
Chymotrypsinogen	Chymotrypsinogen is a precursor of the digestive enzyme chymotrypsin (zymogen.)
	This molecule is inactive and must be cleaved by trypsin, and then by other chymotrypsin molecules before it can reach its full activity. Its activity is the conversion of proteins to amino acids.
Hydrolysis	Hydrolysis is a chemical reaction during which one or more water molecules are split into hydrogen and hydroxide ions, which may go on to participate in further reactions. It is the type of reaction that is used to break down certain polymers, especially those made by step-growth polymerization. Such polymer degradation is usually catalysed by either acid e.g. concentrated sulfuric acid (H_2SO_4) or alkali e.g. sodium hydroxide (NaOH) attack, often increasing with their strength or pH.

Hydrolysis is distinct from hydration, where the hydrated molecule does not "lyse" (break into two new compounds.)

Hemoglobin

Hemoglobin is the iron-containing oxygen-transport metalloprotein in the red blood cells of vertebrates, and the tissues of some invertebrates.

In mammals, the protein makes up about 97% of the red blood cell"s dry content, and around 35% of the total content . Hemoglobin transports oxygen from the lungs or gills to the rest of the body where it releases the oxygen for cell use.

Spectroscopy	Spectroscopy was originally the study of the interaction between radiation and matter as a function of wavelength (λ.) In fact, historically, Spectroscopy referred to the use of visible light dispersed according to its wavelength, e.g. by a prism. Later the concept was expanded greatly to comprise any measurement of a quantity as function of either wavelength or frequency.
Emission	In physics, Emission is the process by which the energy of a photon is released by another entity, for example, by an atom whose electrons make a transition between two electronic energy levels. The emitted energy is in the form of a photon. The emittance of an object quantifies how much light is emitted by it.
Lyman series	In physics and chemistry, the Lyman series is the series of transitions and resulting ultraviolet emission lines of the hydrogen atom as an electron goes from n ≥ 2 to n = 1 (where n is the principal quantum number referring to the energy level of the electron.) The transitions are named sequentially by Greek letters: from n = 2 to n = 1 is called Lyman-alpha, 3 to 1 is Lyman-beta, 4 to 1 is Lyman-gamma, etc. The series is named after its discoverer, Theodore Lyman.
Electron diffraction	Electron diffraction is a technique used to study matter by firing electrons at a sample and observing the resulting interference pattern. This phenomenon occurs due to wave-particle duality, which states that a particle of matter (in this case the incident electron) can be described as a wave. For this reason, an electron can be regarded as a wave much like sound or water waves.
Boundary	In thermodynamics, a Boundary is a real or imaginary volumetric demarcation region drawn around a thermodynamic system across which quantities such as heat, mass, or work can flow. In short, a thermodynamic Boundary is a division between a system and its surroundings. A Boundary may be adiabatic, isothermal, diathermal, insulating, permeable, or semipermeable.
Atomic orbital	An Atomic orbital is a mathematical function that describes the wave-like behavior of either one electron or a pair of electrons, in an atom. This function can be used to calculate the probability of finding any electron of an atom in any specific region around the atom"s nucleus. These functions may serve as three-dimensional graph of an electron"s likely location.
Electron density	Electron density is the measure of the probability of an electron being present at a specific location. In molecules, regions of Electron density are usually found around the atom, and its bonds. In de-localized or conjugated systems, such as phenol, benzene and compounds such as hemoglobin and chlorophyll, the Electron density covers an entire region, i.e., in benzene they are found above and below the planar ring.
Radial distribution function	In computational mechanics and statistical mechanics, a Radial distribution function , g(r), describes how the density of surrounding matter varies as a function of the distance from a particular point.

Suppose, for example, that we choose a molecule at some point O in the volume. What is then the average density at some point P at a distance r away from O? If ρ = N / V is the average density, then the mean density at P given that there is a molecule at O would differ from ρ by some factor g (r.)

Energy levels	A quantum mechanical system or particle that is bound, confined spatially, can only take on certain discrete values of energy, as opposed to classical particles, which can have any energy. These values are called Energy levels. The term is most commonly used for the Energy levels of electrons in atoms or molecules, which are bound by the electric field of the nucleus.
Electron configuration	In atomic physics and quantum chemistry, Electron configuration is the arrangement of electrons of an atom, a molecule, or other physical structure. It concerns the way electrons can be distributed in the orbitals of the given system (atomic or molecular for instance.) Like other elementary particles, the electron is subject to the laws of quantum mechanics, and exhibits both particle-like and wave-like nature.
Ionization	Ionization is the physical process of converting an atom or molecule into an ion by adding or removing charged particles such as electrons or other ions. This is often confused with dissociation (chemistry.) The process works slightly differently depending on whether an ion with a positive or a negative electric charge is being produced.
Electron affinity	The Electron affinity, E_{ea}, of an atom or molecule is the amount of energy required to detach an electron from a singly charged negative ion, i.e., the energy change for the process $$X^- \rightarrow X + e^-$$ An equivalent definition is the energy released ($E_{initial} - E_{final}$) when an electron is attached to a neutral atom or molecule. All elements whose Electron affinity have been measured using modern methods have a positive Electron affinity, but older texts mistakenly report that some elements such as alkaline earth metals have negative E_{ea}, meaning they would repel electrons. This is not recognized by modern chemists.

Absorption	Absorption, in chemistry, is a physical or chemical phenomenon or a process in which atoms, molecules liquid or solid material. This is a different process from adsorption, since the molecules are taken up by the volume, not by surface. A more general term is sorption which covers adsorption, Absorption, and ion exchange.
Dissociation	Dissociation in chemistry and biochemistry is a general process in which ionic compounds (complexes, molecules ions usually in a reversible manner. When a Bronsted-Lowry acid is put in water, a covalent bond between an electronegative atom and a hydrogen atom is broken by heterolytic fission, which gives a proton and a negative ion. Dissociation is the opposite of association and recombination.
Methane	Methane is a chemical compound with the molecular formula CH_4. It is the simplest alkane, and the principal component of natural gas. Methane"s bond angles are 109.5 degrees.
Ethylene	Ethylene is the chemical compound with the formula C_2H_4. It is the simplest alkene. Because it contains a carbon-carbon double bond, Ethylene is called an unsaturated hydrocarbon or an olefin.
Acetylene	Acetylene is the chemical compound with the formula HC_2H. It is a hydrocarbon and the simplest alkyne. This colourless gas is widely used as a fuel and a chemical building block. It is unstable in pure form and thus is usually handled as a solution.
Ammonia	Ammonia is a compound of nitrogen and hydrogen with the formula NH_3. It is normally encountered as a gas with a characteristic pungent odor. Ammonia contributes significantly to the nutritional needs of terrestrial organisms by serving as a precursor to foodstuffs and fertilizers.
Molecular orbital	In chemistry, a Molecular orbital is a mathematical function that describes the wave-like behavior of an electron in a molecule. This function can be used to calculate chemical and physical properties such as the probability of finding an electron in any specific region. The use of the term "orbital" was first used in English by Robert S. Mulliken in 1925 as the English translation of Schrödinger"s use of the German word, "Eigenfunktion".
Molecular orbital theory	In chemistry, Molecular orbital theory is a method for determining molecular structure in which electrons are not assigned to individual bonds between atoms, but are treated as moving under the influence of the nuclei in the whole molecule. In this theory, each molecule has a set of molecular orbitals, in which it is assumed that the molecular orbital wave function ψ_f may be written as a simple weighted sum of the n constituent atomic orbitals χ_i, according to the following equation: $$\psi_j = \sum_{i=1}^{n} c_{ij}\chi_i$$

The c_{ij} coefficients may be determined numerically by substitution of this equation into the Schrödinger equation and application of the variational principle. This method is called the linear combination of atomic orbitals approximation and is used in computational chemistry.

Atomic orbital	An Atomic orbital is a mathematical function that describes the wave-like behavior of either one electron or a pair of electrons, in an atom. This function can be used to calculate the probability of finding any electron of an atom in any specific region around the atom"s nucleus. These functions may serve as three-dimensional graph of an electron"s likely location.
Electron configuration	In atomic physics and quantum chemistry, Electron configuration is the arrangement of electrons of an atom, a molecule, or other physical structure. It concerns the way electrons can be distributed in the orbitals of the given system (atomic or molecular for instance.) Like other elementary particles, the electron is subject to the laws of quantum mechanics, and exhibits both particle-like and wave-like nature.
Order	Order in a crystal lattice is the arrangement of some property with respect to atomic positions. It arises in charge ordering, spin ordering, magnetic ordering, and compositional ordering. It is a thermodynamic entropy concept often displayed by a second Order phase transition.
Fluoride	Fluoride is the anion F^-, the reduced form of fluorine. Both organic and inorganic compounds containing the element fluorine are sometimes called Fluoride s. Fluoride like other halides, is a monovalent ion (–1 charge.)
Energy levels	A quantum mechanical system or particle that is bound, confined spatially, can only take on certain discrete values of energy, as opposed to classical particles, which can have any energy. These values are called Energy levels. The term is most commonly used for the Energy levels of electrons in atoms or molecules, which are bound by the electric field of the nucleus.
Base	In chemistry, a Base is most commonly thought of as an aqueous substance that can accept hydrogen ions..Bases are also the oxides or hydroxides of metals.A soluble Base is also often referred to as an alkali if OH^- ions are involved. This refers to the Brønsted-Lowry theory of acids and bases. Alternate definitions of bases include electron pair donors (Lewis), as sources of hydroxide anions (Arrhenius.)
Acid	An Acid is traditionally considered any chemical compound that, when dissolved in water, gives a solution with a hydrogen ion activity greater than in pure water, i.e. a pH less than 7.0. That approximates the modern definition of Johannes Nicolaus Brønsted and Martin Lowry, who independently defined an Acid as a compound which donates a hydrogen ion (H^+) to another compound (called a base.) Common examples include acetic Acid and sulfuric Acid (used in car batteries.)

Component	In thermodynamics, a Component is a chemically distinct constituent of a system. Calculating the number of components in a system is necessary, for example, when applying Gibbs phase rule in determination of the number of degrees of freedom of a system.
	The number of components is equal to the number of independent chemical constituents, minus the number of chemical reactions between them, minus the number of any constraints (like charge neutrality or balance of molar quantities.)
Carbon monoxide	Carbon monoxide, with the chemical formula CO, is a colorless, odorless and tasteless, yet highly toxic gas. Its molecules consist of one carbon atom and one oxygen atom, connected by a covalent double bond and a dative covalent bond. It is the simplest oxocarbon, and can be viewed as the anhydride of formic acid (CH_2O_2.)
Hemoglobin	Hemoglobin is the iron-containing oxygen-transport metalloprotein in the red blood cells of vertebrates, and the tissues of some invertebrates.
	In mammals, the protein makes up about 97% of the red blood cell"s dry content, and around 35% of the total content . Hemoglobin transports oxygen from the lungs or gills to the rest of the body where it releases the oxygen for cell use.
Metallothionein	Metallothionein is a family of cysteine-rich, low molecular weight (MW ranging from 3500 to 14000 Da) proteins. MTs have the capacity to bind both physiological (Zn, Cu, Se,...) and xenobiotic (Cd, Hg, Ag,...)

Ionic bond	An Ionic bond is a type of chemical bond that involves a metal and a non-metal ion (or polyatomic ions such as ammonium) through electrostatic attraction. In short, it is a bond formed by the attraction between two oppositely charged ions. The metal donates one or more electrons, forming a positively charged ion or cation with a stable electron configuration.
Hemoglobin	Hemoglobin is the iron-containing oxygen-transport metalloprotein in the red blood cells of vertebrates, and the tissues of some invertebrates. In mammals, the protein makes up about 97% of the red blood cell"s dry content, and around 35% of the total content . Hemoglobin transports oxygen from the lungs or gills to the rest of the body where it releases the oxygen for cell use.
Base	In chemistry, a Base is most commonly thought of as an aqueous substance that can accept hydrogen ions..Bases are also the oxides or hydroxides of metals.A soluble Base is also often referred to as an alkali if OH^- ions are involved. This refers to the Brønsted-Lowry theory of acids and bases. Alternate definitions of bases include electron pair donors (Lewis), as sources of hydroxide anions (Arrhenius.)
Liquid	Liquid is one of the principal states of matter. A Liquid is a fluid that has the particles loose and can freely form a distinct surface at the boundaries of its bulk material. The surface is a free surface where the Liquid is not constrained by a container.
Radial distribution function	In computational mechanics and statistical mechanics, a Radial distribution function , g(r), describes how the density of surrounding matter varies as a function of the distance from a particular point. Suppose, for example, that we choose a molecule at some point O in the volume. What is then the average density at some point P at a distance r away from O? If $\rho = N / V$ is the average density, then the mean density at P given that there is a molecule at O would differ from ρ by some factor g (r.)
Surface tension	Surface tension is an attractive property of the surface of a liquid. It is what causes the surface portion of liquid to be attracted to another surface, such as that of another portion of liquid (as in connecting bits of water or as in a drop of mercury that forms a cohesive ball.) Applying Newtonian physics to the forces that arise due to Surface tension accurately predicts many liquid behaviors that are so commonplace that most people take them for granted.
Methane	Methane is a chemical compound with the molecular formula CH_4. It is the simplest alkane, and the principal component of natural gas. Methane"s bond angles are 109.5 degrees.
Hydrate	Hydrate is a term used in inorganic chemistry and organic chemistry to indicate that a substance contains water. The chemical state of the water varies widely between Hydrate s, some of which were so labeled before their chemical structure was understood.

In organic chemistry, a Hydrate is a compound formed by the addition of water or its elements to another molecule.

Absorption	Absorption, in chemistry, is a physical or chemical phenomenon or a process in which atoms, molecules liquid or solid material. This is a different process from adsorption, since the molecules are taken up by the volume, not by surface. A more general term is sorption which covers adsorption, Absorption, and ion exchange.
Absorption spectroscopy	Absorption spectroscopy refers to a range of techniques employing the interaction of electromagnetic radiation with matter. (Spectroscopy is a word that has come to denote an even wider variety of techniques used in physics and chemistry.) In Absorption spectroscopy, the intensity of a beam of light measured before and after interaction with a sample is compared.
Pressure	Pressure is the force per unit area applied in a direction perpendicular to the surface of an object. Gauge Pressure is the Pressure relative to the local atmospheric or ambient Pressure. Pressure is an effect which occurs when a force is applied on a surface.
Resolution	Resolution in terms of electron density is a measure of the resolvability in the electron density map of a molecule. In X-ray crystallography, Resolution is the highest resolvable peak in the diffraction pattern. While cryo-electron microscopy is a frequency space comparison of two halves of the data, which strives to correlate with the X-ray definition.
Emission	In physics, Emission is the process by which the energy of a photon is released by another entity, for example, by an atom whose electrons make a transition between two electronic energy levels. The emitted energy is in the form of a photon. The emittance of an object quantifies how much light is emitted by it.
Multiplicity	Multiplicity in quantum chemistry is used to distinguish between several degenerate wavefunctions that differ only in the orientation of their angular spin momenta. It is defined as 2S+1, where S is the angular spin momentum. Multiplicity is the quantification of the amount of unpaired electron spin.
Absorbance	In spectroscopy, the Absorbance A (also called optical density) is defined as $$A_\lambda = -\log_{10}(I/I_0),$$ where I is the intensity of light at a specified wavelength λ that has passed through a sample (transmitted light intensity) and I_0 is the intensity of the light before it enters the sample or incident light intensity. Absorbance measurements are often carried out in analytical chemistry, since the Absorbance of a sample is proportional to the thickness of the sample and the concentration of the absorbing species in the sample, in contrast to the transmittance I / I_0 of a sample, which varies logarithmically with thickness and concentration.

Outside the field of analytical chemistry, e.g. when used with the Tunable Diode Laser Absorption Spectroscopy (TDLAS) technique, the Absorbance is sometimes defined as the natural logarithm instead of the base-10 logarithm, i.e. as

$$A_\lambda = -\ln(I/I_0)$$

The term absorption refers to the physical process of absorbing light, while Absorbance refers to the mathematical quantity.

Infrared spectroscopy	Infrared spectroscopy is the subset of spectroscopy that deals with the infrared region of the electromagnetic spectrum. It covers a range of techniques, the most common being a form of absorption spectroscopy. As with all spectroscopic techniques, it can be used to identify compounds or investigate sample composition.
Normal mode	A Normal mode of an oscillating system is a pattern of motion in which all parts of the system move sinusoidally with the same frequency. The frequencies of the Normal mode s of a system are known as its natural frequencies or resonant frequencies. A physical object, such as a building, bridge or molecule, has a set of Normal mode s that depend on its structure and composition.
Franck-Condon principle	The Franck-Condon principle is a rule in spectroscopy and quantum chemistry that explains the intensity of vibronic transitions. Vibronic transitions are the simultaneous changes in electronic and vibrational energy levels of a molecule due to the absorption or emission of a photon of the appropriate energy. The principle states that during an electronic transition, a change from one vibrational energy level to another will be more likely to happen if the two vibrational wave functions overlap more significantly.
Spectroscopy	Spectroscopy was originally the study of the interaction between radiation and matter as a function of wavelength (λ.) In fact, historically, Spectroscopy referred to the use of visible light dispersed according to its wavelength, e.g. by a prism. Later the concept was expanded greatly to comprise any measurement of a quantity as function of either wavelength or frequency.
DNA	Deoxyribonucleic acid (DNA) is a nucleic acid that contains the genetic instructions used in the development and functioning of all known living organisms and some viruses. The main role of DNA molecules is the long-term storage of information. DNA is often compared to a set of blueprints or a recipe, or a code, since it contains the instructions needed to construct other components of cells, such as proteins and RNA molecules.
Acid	An Acid is traditionally considered any chemical compound that, when dissolved in water, gives a solution with a hydrogen ion activity greater than in pure water, i.e. a pH less than 7.0. That approximates the modern definition of Johannes Nicolaus Brønsted and Martin Lowry, who independently defined an Acid as a compound which donates a hydrogen ion (H^+) to another compound (called a base.) Common examples include acetic Acid and sulfuric Acid (used in car batteries.)

Melting	Melting is a physical process that results in the phase change of a substance from a solid to a liquid. The internal energy of a solid substance is increased, typically by the application of heat or pressure, resulting in a rise of its temperature to the Melting point, at which the rigid ordering of molecular entities in the solid breaks down to a less-ordered state and the solid liquefies. An object that has melted completely is molten.
Isosbestic point	In spectroscopy, an Isosbestic point is a specific wavelength at which two chemical species have the same molar absorptivity (ε) or -more generally- are linearly related.
	When an isosbestic plot is constructed by the superposition of the absorption spectra of two species (whether by using molar absorptivity for the representation, or by using absorbance and keeping the same molar concentration for both species), the Isosbestic point corresponds to a wavelength at which these spectra cross each other.
	A pair of substances can have several Isosbestic point s in their spectra.
Chemical shift	In nuclear magnetic resonance (NMR), the Chemical shift describes the dependence of nuclear magnetic energy levels on the electronic environment in a molecule. Chemical shift s are relevant in NMR spectroscopy techniques such as proton NMR and carbon-13 NMR.
	An atomic nucleus can have a magnetic moment (nuclear spin), which gives rise to different energy levels and resonance frequencies in a magnetic field. The total magnetic field experienced by a nucleus includes local magnetic fields induced by currents of electrons in the molecular orbitals (note that electrons have a magnetic moment themselves.)
Fourier transform spectroscopy	Fourier transform spectroscopy is a measurement technique whereby spectra are collected based on measurements of the temporal coherence of a radiative source, using time-domain measurements of the electromagnetic radiation or other type of radiation. It can be applied to a variety of types of spectroscopy including optical spectroscopy, infrared spectroscopy (FT IR, FT-NIRS), Fourier transform (FT) nuclear magnetic resonance, mass spectrometry and electron spin resonance spectroscopy. There are several methods for measuring the temporal coherence of the light, including the continuous wave Michelson or Fourier transform spectrometer and the pulsed Fourier transform spectrograph (which is more sensitive and has a much shorter sampling time than conventional spectroscopic techniques, but is only applicable in a laboratory environment.)
Relaxation	In nuclear magnetic resonance (NMR) spectroscopy and magnetic resonance imaging (MRI) the term Relaxation describes several processes by which nuclear magnetization prepared in a non-equilibrium state return to the equilibrium distribution. In other words, Relaxation describes how fast spins "forget" the direction in which they are oriented. The rates of this spin Relaxation can be measured in both spectroscopy and imaging applications.

Chemical kinetics	Chemical kinetics is the study of rates of chemical processes. Chemical kinetics includes investigations of how different experimental conditions can influence the speed of a chemical reaction and yield information about the reaction"s mechanism and transition states, as well as the construction of mathematical models that can describe the characteristics of a chemical reaction. In 1864, Peter Waage and Cato Guldberg pioneered the development of Chemical kinetics by formulating the law of mass action, which states that the speed of a chemical reaction is proportional to the quantity of the reacting substances.
Magnetic resonance imaging	Magnetic resonance imaging is primarily a medical imaging technique most commonly used in radiology to visualize the internal structure and function of the body. Magnetic resonance imaging provides much greater contrast between the different soft tissues of the body than computed tomography (CT) does, making it especially useful in neurological (brain), musculoskeletal, cardiovascular, and oncological (cancer) imaging. Unlike CT, it uses no ionizing radiation, but uses a powerful magnetic field to align the nuclear magnetization of (usually) hydrogen atoms in water in the body.
Fluorescence	Fluorescence is a luminescence that is mostly found as an optical phenomenon in cold bodies, in which the molecular absorption of a photon triggers the emission of a photon with a longer (less energetic) wavelength, though a shorter wavelength emission is sometimes observed from multiple photon absorption. The energy difference between the absorbed and emitted photons ends up as molecular rotations, vibrations or heat. Sometimes the absorbed photon is in the ultraviolet range, and the emitted light is in the visible range, but this depends on the absorbance curve and Stokes shift of the particular fluorophore.
Phosphorescence	Phosphorescence is a specific type of photoluminescence related to fluorescence. Unlike fluorescence, a phosphorescent material does not immediately re-emit the radiation it absorbs. The slower time scales of the re-emission are associated with "forbidden" energy state transitions in quantum mechanics.
Liquid	Liquid is one of the principal states of matter. A Liquid is a fluid that has the particles loose and can freely form a distinct surface at the boundaries of its bulk material. The surface is a free surface where the Liquid is not constrained by a container.
Quantum yield	The Quantum yield of a radiation-induced process is the number of times that a defined event occurs per photon absorbed by the system. Thus, the Quantum yield is a measure of the efficiency with which absorbed light produces some effect.
	For example, in a chemical photodegradation process, when a molecule falls apart after absorbing a light quantum, the Quantum yield is the number of destroyed molecules divided by the number of photons absorbed by the system.
Inversion	In meteorology, an Inversion is a deviation from the normal change of an atmospheric property with altitude. It almost always refers to a temperature Inversion, i.e., an increase in temperature with height, or to the layer (Inversion layer) within which such an increase occurs.

An Inversion can lead to pollution such as smog being trapped close to the ground, with possible adverse effects on health.

Monochromator

A Monochromator is an optical device that transmits a mechanically selectable narrow band of wavelengths of light or other radiation chosen from a wider range of wavelengths available at the input. The name is from the Greek roots mono-, single, and chroma, colour, and the Latin suffix -ator, denoting an agent.

A device that can produce monochromatic light has many uses in science and in optics because many optical characteristics of a material are dependent on color.

Laser-induced fluorescence

Laser-induced fluorescence is a spectroscopic method used for studying structure of molecules, detection of selective species and flow visualization and measurements.

The species to be examined is excited with a laser. The wavelength is often selected to be the one at which the species has its largest cross section.

Quantum yield	The Quantum yield of a radiation-induced process is the number of times that a defined event occurs per photon absorbed by the system. Thus, the Quantum yield is a measure of the efficiency with which absorbed light produces some effect. For example, in a chemical photodegradation process, when a molecule falls apart after absorbing a light quantum, the Quantum yield is the number of destroyed molecules divided by the number of photons absorbed by the system.
Hemoglobin	Hemoglobin is the iron-containing oxygen-transport metalloprotein in the red blood cells of vertebrates, and the tissues of some invertebrates. In mammals, the protein makes up about 97% of the red blood cell"s dry content, and around 35% of the total content . Hemoglobin transports oxygen from the lungs or gills to the rest of the body where it releases the oxygen for cell use.
DNA	Deoxyribonucleic acid (DNA) is a nucleic acid that contains the genetic instructions used in the development and functioning of all known living organisms and some viruses. The main role of DNA molecules is the long-term storage of information. DNA is often compared to a set of blueprints or a recipe, or a code, since it contains the instructions needed to construct other components of cells, such as proteins and RNA molecules.
Melanin	Melanin is a class of compounds found in plants, animals, and protists, where it serves predominantly as a pigment. The class of pigments are derivatives of the amino acid tyrosine. Many Melanin s are insoluble salts and show affinity to water.
Absorption	Absorption, in chemistry, is a physical or chemical phenomenon or a process in which atoms, molecules liquid or solid material. This is a different process from adsorption, since the molecules are taken up by the volume, not by surface. A more general term is sorption which covers adsorption, Absorption, and ion exchange.
Enzymes	Enzymes are biomolecules that catalyze (i.e., increase the rates of) chemical reactions. Nearly all known Enzymes are proteins. However, certain RNA molecules can be effective biocatalysts too.

Boundary	In thermodynamics, a Boundary is a real or imaginary volumetric demarcation region drawn around a thermodynamic system across which quantities such as heat, mass, or work can flow. In short, a thermodynamic Boundary is a division between a system and its surroundings. A Boundary may be adiabatic, isothermal, diathermal, insulating, permeable, or semipermeable.
Kinetic theory	Kinetic theory attempts to explain macroscopic properties of gases, such as pressure, temperature by considering their molecular composition and motion. Essentially, the theory posits that pressure is due not to static repulsion between molecules, as was Isaac Newton"s conjecture, but due to collisions between molecules moving at different velocities. Kinetic theory is also known as the Kinetic-Molecular Theory or the Collision Theory or the Kinetic-Molecular Theory of Gases.
Density	The Density of a material is defined as its mass per unit volume. The symbol of Density is ρ ">rho.) Mathematically: $$\rho = \frac{m}{V}$$ where: ρ is the Density, m is the mass, V is the volume.
Liquid	Liquid is one of the principal states of matter. A Liquid is a fluid that has the particles loose and can freely form a distinct surface at the boundaries of its bulk material. The surface is a free surface where the Liquid is not constrained by a container.
Solution	In chemistry, a Solution is a homogeneous mixture composed of two or more substances. In such a mixture, a solute is dissolved in another substance, known as a solvent. Gases may dissolve in liquids, for example, carbon dioxide or oxygen in water.
Electrophoresis	Electrophoresis is the best-known electrokinetic phenomenon. It was discovered by Reuss in 1807. He observed that clay particles dispersed in water migrate under influence of an applied electric field.
Isoelectric point	The Isoelectric point, sometimes abbreviated to IEP, is the pH at which a particular molecule or surface carries no net electrical charge. Amphoteric molecules called zwitterions contain both positive and negative charges depending on the functional groups present in the molecule. The net charge on the molecule is affected by pH of their surrounding environment and can become more positively or negatively charged due to the loss or gain of protons (H^+.)

Gel	A Gel is a solid, jelly-like material that can have properties ranging from soft and weak to hard and tough. Gel s are defined as a substantially dilute crosslinked system, which exhibits no flow when in the steady-state. By weight, Gel s are mostly liquid, yet they behave like solids due to a three-dimensional crosslinked network within the liquid.
Polyethylene	Polyethylene or polythene (IUPAC name polyethene or poly(methylene)) is a thermoplastic commodity heavily used in consumer products (notably the plastic shopping bag.) Over 60 million tons of the material are produced worldwide every year.
	Polyethylene is a polymer consisting of long chains of the monomer ethylene (IUPAC name ethene.)
Polymer	A Polymer is a large molecule composed of repeating structural units typically connected by covalent chemical bonds. While Polymer in popular usage suggests plastic, the term actually refers to a large class of natural and synthetic materials with a variety of properties.
	Due to the extraordinary range of properties accessible in Polymer ic materials , they have come to play an essential and ubiquitous role in everyday life - from plastics and elastomers on the one hand to natural bio Polymer s such as DNA and proteins that are essential for life on the other.
Beta	The Beta of a plasma, symbolized by β, is the ratio of the plasma pressure ($p = n\ k_B\ T$) to the magnetic pressure ($p_{mag} = B^2/2\mu_0$.)
	$$\beta = \frac{p}{p_{mag}} = \frac{nk_BT}{(B^2/2\mu_0)}$$
	Beta is a parameter indicating the relative importance of kinetic to electromagnetic phenomena. In fusion power applications, Beta can be thought of as an economic figure of merit.
DNA	Deoxyribonucleic acid (DNA) is a nucleic acid that contains the genetic instructions used in the development and functioning of all known living organisms and some viruses. The main role of DNA molecules is the long-term storage of information. DNA is often compared to a set of blueprints or a recipe, or a code, since it contains the instructions needed to construct other components of cells, such as proteins and RNA molecules.
Hemoglobin	Hemoglobin is the iron-containing oxygen-transport metalloprotein in the red blood cells of vertebrates, and the tissues of some invertebrates.
	In mammals, the protein makes up about 97% of the red blood cell"s dry content, and around 35% of the total content . Hemoglobin transports oxygen from the lungs or gills to the rest of the body where it releases the oxygen for cell use.

Melting	Melting is a physical process that results in the phase change of a substance from a solid to a liquid. The internal energy of a solid substance is increased, typically by the application of heat or pressure, resulting in a rise of its temperature to the Melting point, at which the rigid ordering of molecular entities in the solid breaks down to a less-ordered state and the solid liquefies. An object that has melted completely is molten.
Thermodynamic	In physics, Thermodynamic s ">power") is the study of the conversion of energy into work and heat and its relation to macroscopic variables such as temperature,volume and pressure. Its underpinnings, based upon statistical predictions of the collective motion of particles from their microscopic behavior, is the field of statistical Thermodynamic s, a branch of statistical mechanics. Historically, Thermodynamic s developed out of need to increase the efficiency of early steam engines.
Enzymes	Enzymes are biomolecules that catalyze (i.e., increase the rates of) chemical reactions. Nearly all known Enzymes are proteins. However, certain RNA molecules can be effective biocatalysts too.
Exact differential	A mathematical differential dQ is said to be exact, as contrasted with an in Exact differential , if the differentiable function Q exists. However, if dQ is chosen arbitrarily, a corresponding Q might not exist. For one dimension, a differential $$dQ = A(x)dx$$ is always exact.
Inexact Differential	In thermodynamics, an Inexact differential or imperfect differential is any quantity, particularly heat Q and work W, that are not state functions (a property of a system that depends only on the current state of the system, not on the way in which the system acquired that state), in that their values depend on how the process is performed. The symbol , or δ (in the modern sense), which originated from the work of German mathematician Carl Gottfried Neumann in his 1875 Vorlesungen über die mechanische Theorie der Wärme, indicates that Q and W are path dependent. In terms of infinitesimal quantities, the first law of thermodynamics is thus expressed as: $$dU = \delta Q - \delta W$$ where δQ and δW are "inexact", i.e. path-dependent, and dU is "exact", i.e. path-independent.

Enthalpy

In thermodynamics and molecular chemistry, the Enthalpy is a thermodynamic property of a fluid. It can be used to calculate the heat transfer during a quasistatic process taking place in a closed thermodynamic system under constant pressure. Enthalpy H is an arbitrary concept but the Enthalpy change ΔH is more useful because it is equal to the change in the internal energy of the system, plus the work that the system has done on its surroundings.

Entropy

Entropy is a concept applied across physics, information theory, mathematics and other branches of science and engineering. The following definition is shared across all these fields:

$$S = -k \sum_i P_i \ln P_i$$

where S is the conventional symbol for Entropy. The sum runs over all microstates consistent with the given macrostate and P_i is the probability of the ith microstate.

CPSIA information can be obtained at www.ICGtesting.com
Printed in the USA
LVOW020053150911

246286LV00001B/175/P